5G LTE Narrowband Internet of Things (NB-IoT)

Hossam Fattah

5G LTE Narrowband Internet of Things (NB-IoT)

CRC Press
Taylor & Francis Group
Boca Raton London New York

CRC Press is an imprint of the
Taylor & Francis Group, an **Informa** business

CRC Press
Taylor & Francis Group
6000 Broken Sound Parkway NW, Suite 300
Boca Raton, FL 33487-2742

© 2019 by Hossam Fattah.

CRC Press is an imprint of Taylor & Francis Group, an Informa business

No claim to original U.S. Government works

Printed on acid-free paper

International Standard Book Number-13: 978-1-138-31760-4 (Hardback)

Contents

Preface

Broadband wireless communication has evolved dramatically during the past few decades: WiFi, WiMax, ZigBee, Bluetooth, RFID, GSM®, UMTS™, and finally LTE™. These wireless technologies have been readily used by thousands of millions of users around the globe. This is because most of them transformed from a voice-centric technology, to a technology that supports rich multimedia applications, video streaming, rich Internet browsing, chatting, and voice over legacy IP networks.

LTE is the most recent broadband cellular technology aiming at increasing data rates and exploiting advances in digital signal processing such as beam-forming, MIMO, and different channel coding. LTE is now used by many multiple mobile operators around the globe. Existing mobile networks such as legacy GSM, CDMA2000®, and WCDMA are currently converging into LTE due to its efficient modulation scheme, known as OFDM, and other advanced features that enable a new era of applications such as Internet of Things and sidelink communication.

The next phase in LTE advancement is the huge proposal currently under standardization by 3GPP™ (3rd Generation Partnership Project) to transform LTE into a new fifth generation LTE technology aiming at managing different wireless technology; LTE, WiFi, Bluetooth, and legacy cellular standards under one umbrella called 5G. Moreover, the new LTE advancement includes solid support for several applications that have never been supported before in any wireless technology. This includes applications like augmented and virtual reality, Internet of Things, device-to-device communication, machine type communication, carrier aggregation, dual connectivity, relay nodes, autonomous cars, mission-critical applications, industry automation and control, and many others. All of these technologies are bundled under one umbrella starting in Release 15 and targeting what is called *next decade* 5G family of standards that are currently in its initial set of release and standardization by 3GPP and ITU.

Apparently, the LTE family of standards has grown significantly in complexity and features. Although LTE Narrowband Internet of Things (NB-IoT) technology is manifested by low-complexity devices, its protocol stack and operation remain a challenge. The core protocol stack and its operations are defined by 3GPP in many technical specifications and reports which requires careful study, understanding, and browsing of these specifications. Different other protocols, applications, or infrastructures are still needed for the complete operation of NB-IoT in the field.

This book presents the new 5G LTE NB-IoT technology based on Release 15, its protocol stack, architecture, operation, and challenges. The book covers different aspects of NB-IoT and provides necessary details about this technology, while at the same time presents it in simplistic views, as much as possible.

This book is one of few books that aims at providing a single-stop shop for one important feature of 5G LTE. The book focuses on providing comprehensive details about a single 5G LTE feature, which is the cellular NB-IoT. NB-IoT is the next wireless technology to enable a world of digital connectivity in home, city, and building. NB-IoT connected devices are projected to reach a number of hundreds of thousands connected devices per square kilometer.

This book covers the details of 5G LTE NB-IoT across different chapters. In Chapter 2, we introduce the history of LTE, NB-IoT devices, and their requirements and applications. Chapters 3–7 cover the core 3GPP protocol stack of NB-IoT and explain the RRC, PDCP, RLC, MAC, and PHY sublayers. Chapter 8 explains how the quality of service framework is supported in core network for NB-IoT devices. Finally, Chapter 9 covers use cases and deployment recommendations for NB-IoT networks and devices.

Dr. Hossam Fattah
Bellevue, WA, USA

Acknowledgments

The author of the book expresses his gratitude and thanks his family, professional friends, and colleagues for their support, encouragement, and suggestion to produce and publish this book.

3GPP standardization process is a long process. 3GPP technical specifications and reports have been drafted, finalized, and released by many stakeholders and partners including professional individuals, working groups, companies, regulatory bodies, and researchers. We acknowledge their efforts and the final technical specifications that have been released. Without the effort of those stakeholders and their partnership, this book and this technology would not come to the light.

Finally, special thanks to the assistants, editors, and publisher for their continuous effort in improving the materials and presentations of this book.

Notices

Author

Hossam Fattah received his Ph.D. in Electrical and Computer Engineering from University of British Columbia, Vancouver, Canada in 2003. In 2000, he received his master degree in Applied Science in Electrical and Computer Engineering from University of Victoria, Victoria, Canada. He completed his bachelor degree in Computers and Systems Engineering from AlAzhar University, Cairo, Egypt in 1995.

Between 2004 and 2013, he has been with the academia and industry, including Texas A&M University and Spirent Communications, USA, working on wireless communication technology, cellular systems, and research and development for several wireless standards and protocol stacks including Zigbee, WiFi, WiMax, CDMA, and 3G/4G/5G systems.

Since 2013, he has been with Microsoft Corporation, USA, working on different networking products and services for Windows and Cloud Networking. He is also affiliated with University of Washington, Tacoma, USA collaborating on 5G research and innovations.

He has contributed to many technical publications in referred conferences, journal, patents, and talks. He is a registered professional engineer with the Engineers and Geoscientists of British Columbia, Canada.

List of Abbreviations

1G	1st Generation
2G	2nd Generation
3G	3rd Generation
3GPP™	Third Generation Partnership Project
4G	4th Generation
5G	5th Generation
ACK	Acknowledgement
AES	Advanced Encryption Standard
AM	Acknowledgement Mode
AMBR	Aggregate Maximum Bit Rate
AMD	Acknowledged Mode Data
AMPS	Advanced Mobile Phone System
ARP	Allocation and Retention Priority
ARQ	Automatic Repeat Request
AS	Access Stratum
BCCH	Broadcast Common Control Channel
BCH	Broadcast Control Channel
BSR	Buffer Status Report
C-RNTI	Cell Random Network Temporary Identifier
CCCH	Common Control Channel
CDMA	Code Division Multiple Access
CDMA2000-EV-DO	CDMA2000 Evolution Data Optimized
CE	Control Element
CEL	Coverage Enhancement Level

CIoT	Cellular Internet of Things
CMAS	Commercial Mobile Alert Service
CP	Cyclic Prefix
CRC	Cyclic Redundancy Check
CRS	Cell-specific Reference Signal
CSFB	Circuit Switched FallBack
CSG	Closed Subscriber Group
DCCH	Dedicated Control Channel
DFT	Discrete Fourier Transform
DL-SCH	Downlink Shared Channel
DRB	Data Radio Bearer
DRX	Discontinuous Reception
DTCH	Dedicated Traffic Control Channel
E-RAB	EPS Radio Access Bearer
E-UTRA	Evolved UMTS™ Terrestrial Radio Access
E-UTRAN	Evolved UMTS Terrestrial Radio Access Network
EARFCN	E-UTRA Absolute Radio Frequency Channel Number
EDGE	Enhanced Data GSM® Environment
eMBB	Enhanced Mobile BroadBand
EPC	Evolved Packet Core
EPRE	Energy Per Resource Element
ETWS	Earthquake and Tsunami Warning System
FDD	Frequency Division Duplex
FDMA	Frequency Division Multiple Access
FFT	Fast Fourier Transform
GBR	Guaranteed Bit Rate
GSM®	Global System for Mobile Communication
H-SFN	Hyper System Frame Number
HARQ	Hybrid Automatic Repeat reQuest
HFN	Hyper Frame Number
HSS	Home Subscriber Service
IFFT	Inverse Fast Fourier Transform
IMEI	International Mobile Equipment Identity
IMS	IP Multimedia Subsystem

IMSI	International Mobile Subscriber Identity
IMT	International Mobile Telecommunication
IoT	Internet of Things
IS-95	Interim Standard 95
ITU	International Telecommunication Union
LCG	Logical Channel Group
LSB	Least Significant Bit
LTE™	Long Term Evolution
MAC	Medium Access Control Sublayer
MAC	Message Authentication Code
MBMS	Multimedia Broadcast Multicast Service
MBR	Maximum Bit Rate
MCC	Mobile Country Code
MCS	Modulation and Coding Scheme
MIB	Master Information Block
MME	Mobility Management Entity
MNC	Mobile Network Code
MSB	Most Significant Bit
MTC	Machine Type Communication
NACK	Negative Acknowledgment
NAS	Non Access Stratum
NPBCCH	Narrowband Physical Broadcast Control Channel
NPDCCH	Narrowband Physical Downlink Control Channel
NPDSCH	Narrowband Physical Downlink Shared Channel
NPRACH	Narrowband Physical Random Channel
NPUSCH	Narrowband Physical Uplink Shared Channel
NR	New Radio
NRS	Narrowband Reference Signal
OFDM	Orthogonal Frequency Division Multiplexing
P-GW	Packet GateWay
P-RNTI	Paging Random Network Temporary Identifier
PCCH	Paging Common Control Channel

PCH	Paging Control Channel
PCI	Physical Cell ID
PDCP	Packet Data Convergence Protocol Sub-layer
PDN	Packet Data Network
PDU	Protocol Data Unit
PLMN	Public Land Mobile Network (i.e., Mobile Operator)
PRB	Physical Resource Block
QCI	QoS Class Identifier
RA	Random Access
RACH	Random Access Channel
RAT	Radio Access Technology
RB	Radio Bearer
RF	Radio Frame
RLC	Radio Link Control Sublayer
RoHC	Robust Header Compression
RRC	Radio Resource Control Sublayer
RSRP	Reference Signal Received Power
RSRQ	Reference Signal Received Quality
RTT	Round Trip Time
RU	Resource Unit
S-GW	Serving GateWay
S-TMSI	SAE-Temporary Mobile Station Identifier
SC-FDMA	Single Carrier Frequency Division Multiple Access
SC-MCCH	Single Cell Multicast Control Channel
SC-MTCH	Single Cell Multicast Transport Channel
SC-PTM	Single Cell Point To Multipoint
SDU	Service Data Unit
SF	SubFrame
SFN	System Frame Number
SI	System Information
SIB	System Information Block
SINR	Signal-to-Interference-plus-Noise Ratio
SMS	Short Message Service
SRB	Signalling Radio Bearer
TDD	Time Division Duplex
TDMA	Time Frequency Division Multiplexing

TM	Transparent Mode
TMD	Transparent Mode Data
TR	Technical Report
TS	Technical Specification
TTI	Transmission Time Interval
UE	User Equipment
UL-SCH	Uplink Shared Channel
UMD	Unacknowledged Mode Data
UMTS™	Universal Mobile Telecommunications System
USIM	UMTS Subscriber Identity Module
WCDMA	Wideband Code Division Multiple Access
WLAN	Wireless Local Area Network

Chapter 1

Internet of Things

1.1 Introduction

Wireless communication systems and devices have emerged and evolved during the past few decades. Figure 1.1 shows the evolution of wireless technology and networks from 1G technology to today's 4G systems. In the 1980s, wireless networks started as voice-centric communication systems such as 1G and 2G networks represented by AMPS and IS-95 cellular systems. Improvements to 1G and 2G systems were added to support data-centric devices with low-to-medium data rates (e.g., 9.6 Kbps) such as EDGE. 3GPP™ was formed for standardizing and introducing the 3G wireless network known as WCDMA. In 1999, 3G systems, WCDMA and cdma2000, were introduced to support a mix of voice, video, and data services. WCDMA was released by 3GPP in different releases; starting from release 99 up to release 7, which allows true mobile broadband connectivity supporting few Mbits per second.

Release 8 was the start of the 4G activity by 3GPP known as LTE™. 3GPP laid out the new specifications of LTE extending the IMT-2000 specifications [1]. LTE saw its first commercial deployment in 2009. LTE introduces many advanced features compared to its predecessors and offers several advantages such as high speed, low latency, higher spectrum efficiency, simplified and all-IP core network, air interface based on Orthogonal Frequency Division Multiple (OFDM) access, and higher cell capacity. LTE has also been introduced in several releases starting from Release 8 until Release 15. The latter release is the mark of the beginning of 5G activities and standardization.

1

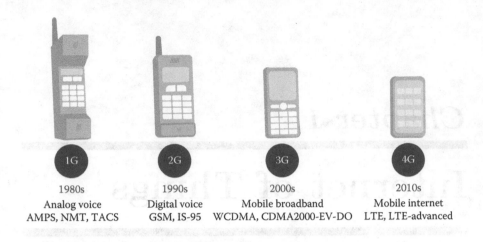

1G	2G	3G	4G
1980s	1990s	2000s	2010s
Analog voice	Digital voice	Mobile broadband	Mobile internet
AMPS, NMT, TACS	GSM, IS-95	WCDMA, CDMA2000-EV-DO	LTE, LTE-advanced

Figure 1.1: Emergence of wireless and cellular networks.

LTE has introduced Machine Type Communication (MTC). MTC is a technology that enables communication between devices and the underlying data transport infrastructure. The data communication can occur between an MTC device and a server, or directly between two MTC devices. MTC manifests itself in a wide range of applications and services. Those applications can be found in different industries, such as healthcare, manufacturing, process automation, energy, and utilities. MTC devices support different network technologies such as point-to-point, multi-hop, ad-hoc networks, or mesh wireless networks. Internet of Things (IoT) is one realization of MTC. MTC devices can be low-complexity, long-range, low-power, or broadband devices. All these devices are communicating with each other and with servers and applications residing on the network.

For example, connected devices used in cars and trucks are characterized by a small data rate while performing a lot of handovers. Smart meters, such as gas or electricity, are stationary and require only a small amount of data which is not delay sensitive. In addition, the number of these MTC devices can be large; ranging from up to few per household to hundreds of thousands per square kilometer. MTC devices are often battery-powered and without any other external power source. The number of connections by these devices are expected to be ultra-large with an estimated active connection density of 200,000 per square kilometer and a device density of 1 million devices per square kilometer.

3GPP NB-IoT, known as LTE Narrowband Internet of Things (NB-IoT), is one category of the MTC that is introduced in LTE starting from Release 13. LTE NB-IoT delivers different levels of optimizations

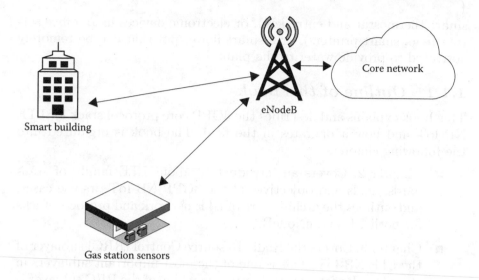

Figure 1.2: Internet of Things applications in smart building and meters.

for NB-IoT devices such as protocol stack and radio interface optimized for NB-IoT, low power consumption, low data rate, no mobility support, limited bandwidth of 180 KHz, extended coverage, and low hardware cost. Figure 1.2 shows one application of NB-IoT in "smart buildings" and smart metering where NB-IoT devices collect a large amount of data and information and send them to a remote server for processing.

The new LTE NB-IoT standard introduced by 3GPP is a stripped version of the full-fledged LTE system in order to keep it as simple as possible while meeting the goals of low cost, minimal power consumption, and extended battery lifetime. NB-IoT devices can be sensors, actuators, wearables, and cameras which form a large number of connected devices or connected "things" such as in smart buildings and sensors in a gas station. These devices are characterized by a non-time critical data transfer and can range from very simple devices to very complex ones. NB-IoT devices connect with the network directly (e.g., eNodeB or Base-Station) through the cellular infrastructure as shown in Figure 1.2.

Devices can range from simple wearables, such as a smart watch or a set of sensors embedded in clothing, to more sophisticated wearable devices monitoring body vital statistics (e.g., heart rate, blood pressure). They can also be non-wearable devices that communicate in a Personal Area Network (PAN) such as a set of home appliances (e.g.,

smart thermostat and entry key), or electronic devices in an office setting (e.g., smart printers), or a smart flower pot that can be remotely
activated to provide water to the plant.

1.1.1 Outline of this book

This book explains and describes the 3GPP core protocol stack for LTE
NB-IoT and how it operates in the field. The book is organized into
the following chapters:

- Chapter 2: Covers an introduction about LTE family of standards, goals and objectives of the 3GPP NB-IoT, its use cases,
and outlines the architecture of LTE network and protocol stacks
for both UE and eNodeB.

- Chapter 3: Covers the Radio Resource Control (RRC) sublayer of
the LTE NB-IoT. This is one of the most important sublayers in
the NB-IoT protocol stack. State machine of the RRC sublayer is
explained. Procedures used to exchange signalling messages with
the eNodeB are presented. This chapter also covers procedures
and behavior of the UE while the UE is idle and not connected
to eNodeB.

- Chapter 4: Covers the Packet Data Convergence Protocol
(PDCP) sublayer. It includes the PDCP sublayer architecture,
procedures for ciphering incoming and outgoing messages to
eNodeB, integrity protection and verification procedures, and
its message format.

- Chapter 5: Covers the Radio Link Control (RLC) sublayer. It
covers how this sublayer provides guaranteed delivery of signalling and data messages to eNodeB. It also includes different
modes of operation and the message format used in each mode.

- Chapter 6: Covers the Medium Access Control (MAC) sublayer.
It explains the random access procedure, contention resolution
procedure, multiplexing and demultiplexing of MAC data units,
and MAC procedures for scheduling uplink traffic.

- Chapter 7: Covers the Physical (PHY) sublayer. It explains the
over-the-air frame structure and traffic exchanges with eNodeB,
uplink and downlink physical channels, physical channels modulation and demodulation, error correction, and procedures used
for transmitting and receiving signalling and data messages.

- Chapter 8: Covers the Quality of Service (QoS) framework and
architecture used by the UE and network for enforcing guaranteed QoS parameters. It explains the end-to-end bearers, QoS

metrics offered by the core network, and QoS guarantees between the UE and the Internet and applications.

■ Chapter 9: Covers the major use cases of NB-IoT applications such as smart parking, smart city, and smart home. It also covers the features to be supported by NB-IoT devices and networks alike for successful deployment and adoption of the NB-IoT technology. It presents both device and network requirements that are planned to be offered by mobile operators releasing the first set of this technology.

1.1.2 How to read this book

This book presents the new 5G 3GPP LTE NB-IoT technology as recently released by 3GPP technical specifications in Release 15. 3GPP technical specifications and technical reports include many other new features of LTE, such as the legacy cellular LTE, relay nodes, or sidelink communication, bundled with the NB-IoT specifications. The new NB-IoT technology is complex and requires a skilled professional in the field of wireless communications and protocols. In this edition of the book, we walk through this new technology, explain its aspects, operations and procedures, control and data paths, and how it is used in different use cases and in every day scenarios. The book contains a lot of illustrations to help in digesting and understanding the technology in a simplistic view.

The book starts with an overview about networking protocol stack terminology and architecture. In the following chapters, each sublayer is presented. Each chapter explains a different sublayer of the NB-IoT technology. Each chapter is also a standalone and self-contained description of the all functionalities, procedures, or data units of the corresponding sublayer.

RRC sublayer includes different procedures and messages. RRC procedures in 3GPP technical specifications are sophisticated procedures and this book focuses and presents the mainstream scenarios of these procedures. The mainstream scenario is the scenario that does not include exceptional behavior, timer expiry events, or abnormal error conditions. This is to provide a simplistic view of each procedure and understand its fundamental operation. RRC messages are released as ASN.1 format by 3GPP. In this book, only relevant fields and important parameters of RRC messages are presented in tables, with their possible values, sizes, or meanings.

PDCP, RLC, and MAC sublayers are also presented in their corresponding chapters. Each of these chapters contains tables about the

Table 1.1   Mathematical Notations Used Throughout This Book

Notation	Meaning
$[x\ y]$	Range of numbers from x to y inclusive
$[x\ y[$	Range of numbers from x to y inclusive of x and exclusive of y
$x \times y$	Product of x and y
$z = x \bmod y$	Module operation. A division of x by y yields integer quotient, q, and integer remainder, z, such that $y \times q + z = x$
$x = y^*$	x is the complex conjugate of y. If $y = a + jb$ then $x = a - jb$
$\begin{bmatrix} x \\ y \\ z \\ w \end{bmatrix}$	Matrix or vector

signalled RRC parameters and configurations to these sublayers which are needed to know how the procedures and functionalities of a sublayer is performed. PHY sublayer is a major sublayer in NB-IoT, and it has been explained in its chapter with great detail.

Finally, NB-IoT use cases, scenarios, features used for NB-IoT deployment, and an application layer protocol suitable for NB-IoT application development are presented. This concludes the book by providing a compressive description about NB-IoT technology starting from the bottom wireless channel and up to the top application level.

Throughout the book, some mathematical notations are used. Those notations are summarized in Table 1.1.

This edition of the book is one of its kind in the area of NB-IoT technology and serves as a handbook for people who need to dig deeper into this technology or are looking for guidance about this new technology. The book presents the most recent and up-to-date information and specifications about NB-IoT. The book is a valuable material for technical and nontechnical readers who are willing to learn and find comprehensive information about the NB-IoT technology.

In future editions of this book, we will cover NB-IoT using different wireless interfaces (e.g., GSM®, WiFi). We will also cover data analytics, data science, and Cloud-based systems used for storing, processing, and analyzing NB-IoT data and information. We wish that the reader will enjoy reading this book and keep it among his/her library.

Chapter 2

4G and 5G Systems

2.1 LTE History

4G cellular technology, known as E-UTRA or LTE™ [2], has been introduced in 3GPP™ Release 8 in 2008 as the broadband cellular technology that exceeds IMT-2000 requirements [1]. 4G comes with advanced capabilities and features such as higher peak data rates (300 Mbps on DL and 75 Mbps on UL), improved system capacity and coverage, better spectrum efficiency, low latency, reduced operating costs, multi-antenna support, flexible bandwidth operation, and seamless integration with existing systems. LTE Release 10 (known as LTE-Advanced) was later introduced as the technology that meets IMT-Advanced requirements [3]. LTE-Advanced significantly enhances the LTE Release 8 by supporting bandwidth extension up to 100 MHz through carrier aggregation to support much higher peak rates (1 Gbps in DL and 500 Mbps in UL), higher throughput and coverage, and lower latencies, resulting in a better user experience. In addition, LTE Release 10 supports a higher number of spatial multiplexing (MIMO), coordinated multi-point transmission, and relay nodes. LTE Release 13 (LTE-Advanced-Pro), released in 2016, extends LTE-Advanced to a wide spectrum of new applications and industries which enable new use cases beyond smartphones. Release 13 was the start of pre-5G activities aiming at complementing 5G new services and features.

5G is designed to support the ITU requirements for IMT-2020 capabilities [4] as shown in Figure 2.1. The peak data rate of IMT-2020 for enhanced mobile broadband is expected to reach 10 Gbps and can increase up to 20 Gbps. The spectrum efficiency is expected to be three

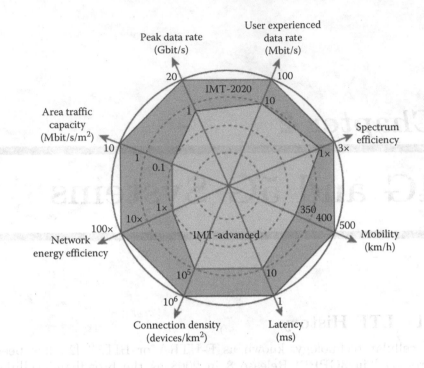

Figure 2.1: ITU target for IMT-2020 and 5G capability [4].[1]

times higher compared to IMT-Advanced. IMT-2020 is expected to support 10 Mbit/s/m² area traffic capacity as the case in hot spots. IMT-2020 would be able to provide 1 ms over-the-air latency, capable of supporting services with very low latency requirements. IMT-2020 is also expected to enable high mobility up to 500 km/h with acceptable Quality of Service (QoS) for high speed trains. Finally, IMT-2020 is expected to support a connection density of up to 10⁶ devices per km², for example, in massive machine type communication scenarios.

5G wireless communication network is the next generation connectivity and technology for the next decade and beyond that is set up to meet the IMT-2020 requirements [4]. 5G LTE standardization and specifications have started in 3GPP Release 15. The initial set of 5G technical specifications, based on LTE and New Radio (NR), was finalized in early 2018 as part of Release 15. The first phase of 5G technical specifications are expected to be fully completed by 2018 and ready for commercial deployment by 2020.

---

[1] Reproduced with a written permission from ITU.

5G technology promises a large number of state-of-the-art features including Narrowband Internet of Things (NB-IoT)[2] for a connected world [5, 6]. 5G is delivering a rich number of features such as connected cars (Vehicle-to-Everything (V2X)), machine type communication, device-to-device communication, small cells, and relay networks. In addition, there are many other advanced features including massive MIMO and advanced antenna techniques, adaptive beamforming, simultaneous use of licensed and unlicensed bands, unified and single air interface, flexible FDD/TDD subframe design, and scalable OFDM numerology and modulation schemes.

While 4G LTE will continue to advance before 5G becomes commercially available, ubiquitous, next generation 5G networks will support many new use cases and vertical applications that simply are not feasible to run over even the most advanced 4G LTE networks. From a variety of IoT deployments to massive machine type communication scenarios, 5G networks will be capable of much more than the mobile broadband applications we have today. These networks will scale to accommodate billions of devices at very high data rates (upwards of 20 Gbps) and ultra-low latency (less than 1 ms) and ultra-high reliability [7].

2.2   5G Narrowband Internet of Things

3GPP 5G technology introduced a new radio interface, the Narrowband IoT (NB-IoT) [5], in Release 13 and has been extended in Release 14 and Release 15. NB-IoT is designed to connect a large number of devices in a wide range of application domains forming so-called Internet of Things (IoT). Connected devices are to communicate through cellular infrastructure. 3GPP has also introduced different data rates suitable for NB-IoT that range from 10 s of Kbps in 180 KHz bandwidth (LTE Cat-NB1) up to few hundreds of Kbps (LTE Cat-NB2) [2, 8]. 5G new radio for NB-IoT is also planned to introduce advanced features for massive IoT including Resource Spread Multiple Access (RSMA) for IoT use cases requiring asynchronous and grant-less access, multi-hop mesh, power saving mode (PSM) schemes, and extended discontinuous reception (eDRX) for longer battery lifetime.

NB-IoT is a low power Wide Area Network (WAN) solution that operates in licensed spectrum bands. 3GPP includes this technology as a part of LTE standards to benefit from the big ecosystem offered by LTE technology and mobile operators.

---

[2]The terms "NB-IoT," "CIoT," "LTE IoT," or "UE" are used interchangeably.

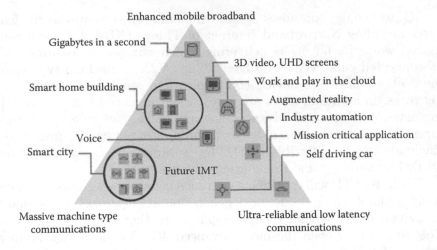

Figure 2.2: Usage scenarios for IMT-2020 and beyond [4].[3]

Not only does 5G technology enhance existing cellular use cases but also expands to a new era of use cases and scenarios; massive IoTs, smart homes, smart cities, smart transportation, smart grids, smart utilities and meters, wearables and remote sensors, autonomous and self-driving vehicles, object tracking, mobile virtual reality, remote control and process automation for aviation and robotics, and mission-critical control [9, 10]. Figure 2.2 illustrates examples of envisioned usage scenarios for IMT-2020 and beyond as set by ITU [4].

It is projected that, over the coming few years, there will be an explosion in the number of IoT connected devices. For example, by 2025, it is expected that more than 5 billion devices will be connected through 5G NB-IoT. 5G NB-IoT devices are designed with the following requirements and goals:

■ **Massive Number of Low-Throughput Devices**: Support at least 52,547 connected devices within a cell site sector. This target was based on using 40 devices per household with the household density based on the assumption for London city provided in [11, 12] (1517 household density per km^2 and cell inter site distance of 1732 m).

■ **Low Power Consumption**: Enable IoT devices to draw low current (in the range of nanoamp) to enable a single battery charge for multiple number of years (in the range of 10 years).

---
[3]Reproduced with a written permission from ITU.

- **Longer Battery Lifetime**: The target is to provide battery life of 10 years with battery capacity of 5 WH.

- **Improved Indoor and Outdoor Coverage**: The target is to achieve an extended coverage of 20 dB compared to legacy GPRS devices. Data rate of at least 160 bps should be supported for both the uplink and downlink.

- **Low Complexity**: The goal is to provide ultra-low complexity devices to support IoT applications that results in a cheaper cost.

- **Low Latency**: A latency of 10 s or less is the target for 99% of the devices.

- **Low Cost**: A target cost of $5 USD per device.

NB-IoT devices are connected to cellular infrastructure and network. Cellular networks, supporting NB-IoT devices, are designed with the following requirements and goals:

- Re-use existing power saving procedures in core network for increasing UE battery lifetime.

- Support sharing the core network between multiple mobile operators.

- Control the UE access for each PLMN. That is, support access class barring per a PLMN.

- Support for Short Message Service (SMS).

- Support IP header compression for IP-based services.

- Support cell selection and (Re)selection procedures in both IDLE and CONNECTED modes.

- Support multicast traffic.

2.3 NB-IoT Applications and Scenarios

Many of the NB-IoT deployment scenarios will be using sensors. Sensors are becoming the endpoints for NB-IoT networks, collecting increasing amount of context aware data and information (e.g., location, images, weather conditions) and injecting a large amount of structured and unstructured data into the networks and applications. Big data, analytics, and predictions have thus become an apparent synonymous for NB-IoT. Those NB-IoT devices used as sensors can be used for the following applications:

- Metering consumption of gas, water, and electricity.

- Measuring weather condition such as temperature, humidity, pressure, wind direction, and Ultra-Violet (UV) index.

- Measuring pollution levels such as carbon emissions, mercury, and radioactive emissions.

- Measuring environmental activities such as noise, pollen and dust levels, and solar activities.

NB-IoT devices can also be used as actuators. Actuators are used to control and steer devices such as controlling traffic lights, traffic lanes, or home appliances. NB-IoT devices used as sensors are typically more in number than those NB-IoT devices acting as actuators.

NB-IoT promises to create a highly connected world that demands using sensors and data analytics for sensing, monitoring, and controlling all events in homes, cars, agricultural, industrial, and environmental venues. Sensors and data generated by sensors are transported and delivered by NB-IoT devices which ultimately facilitate stakeholders to analyze and apply insights in real time. The following examples are now possible with the 5G NB-IoT:

- **Smart City**: Monitoring highway traffic lights and street intersections, monitoring and control of infrastructure grids such as electricity, gas, and sewage; public safety and disaster management; video surveillance; traffic violations; and law enforcement.

- **Smart Home**: Lighting systems, smart appliances, connected TV sets, gaming consoles, sound and theatre systems, smoke and alarm systems, wearables, and kids and pets monitoring devices.

- **Smart Transportation**: Communicating between vehicles, pedestrians, or cyclists for traffic warning, collisions, and accident avoidance, traffic safety and traffic sign enforcement, public buses, trains, and underground transportation information and management, and public parking and parking meter communication.

3GPP defines a number of applications that are typically used by NB-IoT devices and sensors as shown in Table 2.1. These applications are characterized by how much data in bytes are reported continuously for an interval of time [7].

Table 2.1 Summary of CIoT Application and Their Traffic Usage

Application	Number of Devices in a Single Cell	Reporting Interval in Uplink	Number of Uplink Bytes	Total Daily Uplink Traffic (KB)	Reporting Interval in Downlink	Number of Downlink Bytes	Total Daily Downlink Traffic (KB)
Water metering	37500	1/day	200	7324	1/week	50	262
Gas metering	37500	4/hour	100	351652	1/week	50	262
Waste management	100	1/hour	50	117	None	None	0
Pollution monitoring	150	1/hour	1000	3515	2/day	1000	293
Pollution alerting	20	4/hour	5000	9375	1/week	1000	3
Public lighting	200	1/day	20000	3906	2/day	1000	390
Parking management	80000	1/hour	100	187500	1/day	100	7812
Watering	200	2/day	100	39	1/day	100	20
Self-service bike renting	500	4/hour	50	2344	1/hour	50	586
Total	156170			565772			9628

2.4   Massive Number of Low-Throughput Devices

NB-IoT devices are expected in large numbers in home, car, city, and municipality [13, 14]. Two cities, London and Tokyo, are used as a model to know the population and household density, and the number of NB-IoT devices used [11]. The cell geometry is defined as shown in Figure 2.3.

Each cell is designed with a maximum of 40 devices per household. Table 2.2 shows the cell geometry and IoT device density for both London and Tokyo urban areas. The number of devices per cell site sector equals to the area of cell site sector × Household density per km² × number of devices per household.

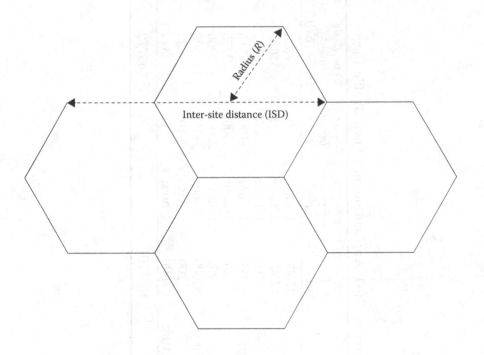

Figure 2.3: CIoT cell geometry.

Table 2.2   IoT Cell Capacity

City	Household Density per km²	ISD (m)	Area of Cell Site Sector (km²)	Number of device per house hold	Number of devices within a Cell Site Sector
London	1517	1732	0.866	40	52548
Tokyo	2316	1732	0.866	40	80226

2.5 Longer Battery Lifetime

NB-IoT devices are powered by a battery. NB-IoT data rate, usage, and coverage scenario determines how long this battery can run with a single charge. When the device is actively transmitting to and receiving from eNodeB, it is consuming battery energy; if the device is in sleep mode, energy consumption is greatly reduced.

Battery capacity is expressed as watt-hour (WH). That is, how much energy, Watt = Volt × Current, it can provide for an hour. Battery lifetime can be approximately calculated by adding the power consumed during the time of being active and the power consumed during which the device is in sleep mode. More power is consumed while the device is active. For this reason, the device is placed in sleep mode for as long as possible to increase the battery lifetime.

Power consumption can vary during active mode. The device might fluctuate between transmitting high volume of traffic or low volume of traffic depending on the traffic load scenario. Coverage can also affect the power consumption as in a poor coverage scenario; the devices needs to transmit and retransmit the same message multiple times resulting in a higher power consumption. For good coverage, the device needs to transmit its messages a fewer number of times that results in a lower power consumption.

The NB-IoT device is constrained in resource, i.e., low-complexity hardware, limited memory and processing power, and no permanent energy source. Different powering mechanisms are available for NB-IoT devices, such as the following:

- Nonrechargeable battery that is charged only once.

- Rechargeable battery with regular recharging (e.g., solar source).

- Rechargeable battery with irregular recharging (e.g., opportunistic energy scavenging).

- Always on (e.g., powered electricity meter).

2.6 Low Latency and Data Reporting

Most of applications running on NB-IoT devices are expected to tolerate delay. That is, they are delay tolerant. Yet, some applications, such as alarming applications, can deliver their data in near real time with a latency target of 10 s.

NB-IoT devices are expected to send and receive data on uplink and from downlink. Such data takes the form of triggered reports, exception reports, or periodic reports [11, 12] as shown in Figure 2.4.

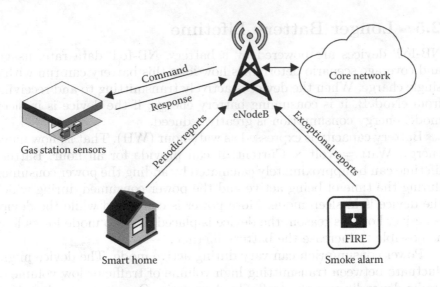

Figure 2.4: Different reports transmitted by CIoT.

In triggered reporting, also known as command-response traffic exchange, eNodeB sends a command to the NB-IoT device where a response from the NB-IoT is optional. Example of commands are such as to turn on/off light or electric switch or to report meter reading. Typically, commands have a payload size of about 20 bytes, whereas response payload size is about 100 bytes with a total roundtrip latency of 10 seconds. Each command-response exchange can be repeated in an interval of 1 day, 2 h, 1 h, or 30 min.

Exception reports are exceptional data transmitted by NB-IoT devices. NB-IoT devices used as a sensor are typically used to monitor a physical condition and if this condition is met, it triggers transmitting an exceptional report. For example, smoke detectors and alarms, gas or power fault detectors, or smart meters transmit exceptional reports which have an uplink payload of 20 bytes and with a latency up to 10 s. Exceptional reports are acknowledged by receiving an ACK signal from eNodeB which has a zero-length packet.

Periodic reports are transmitted parodically by NB-IoT devices. NB-IoT devices used in smart utility (water, gas, or electricity), smart agriculture, or smart environment are examples of those devices that emit periodic reporting. Data size emitted by periodic reporting ranges between 20 bytes and a maximum of 200 bytes. The periodic data is transmitted every interval of time which can be 1 day, 2 h, 1 h, or 30 min.

Finally, software upgrades and updates to NB-IoT devices are expected to occur occasionally. This includes reconfiguration of NB-IoT software or application. Payload of such upgrades is expected to be in the range of 200–2,000 bytes with periodic time interval of 180 days.

Table 2.1 shows different applications that can trigger command-response, exception, or periodic reporting traffic on NB-IoT devices [7].

2.7 LTE NB-IoT Protocol Stack and Architecture

Network protocol stack is designed into a layered architecture that exists at both transmitting and receiving node. Each layer runs a protocol that can communicate with the peer node at the same layer. The protocol exchanges messages, packets, or Protocol Data Units (PDUs) in order to provide services or functions to the upper layer. The protocol also exchanges these messages, packets, or PDUs with the lower layer to use its functions and services.

Figure 2.5 shows the layered architecture. The layers are those of the reference model of Open Systems Interconnection (OSI) which is developed as an international standard for computer networks by the International Standards Organization (ISO). Typically, the layered architecture is further divided vertically into two planes: data plane and control plane. The first is the plane where user data flows between the two nodes, while the latter is where the control information is exchanged. Some of these layers might not exist in the control-plane such as application, session, and presentation layers. Figures 2.7 and 2.8 show the data plane and control plane, respectively, for the NB-IoT protocol stack.

The most bottom two layers in Figure 2.5 are also called the Access Stratum (AS). These two layers are responsible for the handling and processing the physical transmission or reception on the media. In a typical network, the physical media can be an Ethernet cable, wireless channel, or any other form of physical connection. As the media changes from a network to another (for example, from Ethernet to WiFi networks), the MAC and PHY layers protocols need to be changed as well since they are to handle different type of media. In the case of NB-IoT, the physical media is the wireless channel, and both the MAC and PHY layers are referred to as the access stratum. The upper five layers are thus the Non-Access Stratum (NAS) and they are almost the

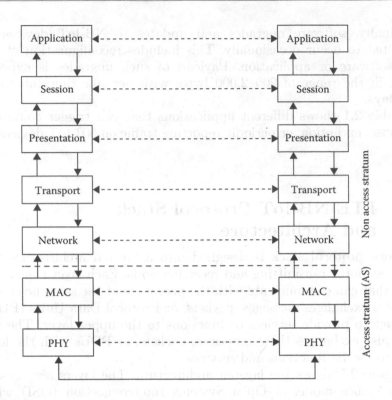

Figure 2.5: OSI data plane protocol stack.

same across different type of physical media since their protocols and functions are independent of the physical media.

Layers shown in Figure 2.5 exchange their data units as Service Data Unit (SDU) or Protocol Data Unit (PDU). SDU refers to the data unit within the layer (i.e., intra-layer data unit), while PDU refers to the data unit exchanged between layers (i.e., inter-layer data unit). SDU and PDU for three layers are shown in Figure 2.6. Each layer has its own SDU which is appended by a header for each layer (H1, H2, H3) when exchanging the SDU to an upper or a lower layer. On the transmit path, each layer appends its SDU by a header and sends it down to the layer below it. On the receive path, each layer sends its SDU to upper layer. Each layer knows the size of its header and thus can strip the PDU received from lower layer to extract the SDU. In addition, each layer may add a trailer to its SDU (e.g., checksum or integrity-protection trailer).

NB-IoT, on the other hand, has the layering architecture of its protocol stack, services, and functions that are to be transmitted and received

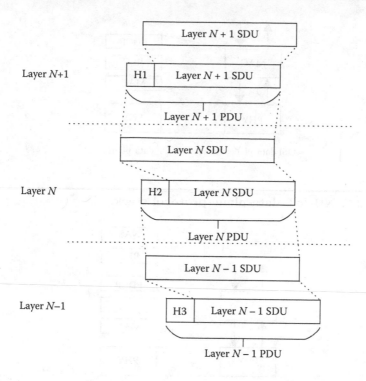

Figure 2.6: Service data unit and protocol data unit.

on a specific type of media; in this case, it is the wireless channel. NB-IoT does not have all layers as in Figure 2.5 defined by 3GPP, but only the most two bottom layers, MAC and PHY layers, while keeping the remaining upper five layers unchanged. This is because 3GPP protocol stack only defines the access stratum and air access methods and protocols which reside only at the MAC and PHY layers. Transport and networking layers protocols (TCP/IP) can be used as they exist today on top of the 3GPP protocols and layers.

Figures 2.7 and 2.8 illustrate the 3GPP protocol stack within the ISO stack for both data plane and control plane. In Figure 2.7 and as expected, 3GPP defines only the access stratum layers: Packet Data Convergence Protocol (PDCP), Radio Link Control (RLC), Medium Access Control (MAC), and Physical (PHY) sublayers, while in Figure 2.8, additional control-plane sublayers are also defined by 3GPP: Radio Resource Control (RRC) and Non-Access Stratum (NAS).

In this book, we describe and explain the 3GPP data-plane and control-plane protocol stack at the PDCP, RLC, MAC, and PHY sublayers. In addition, the RRC sublayer from the control-plane stack is

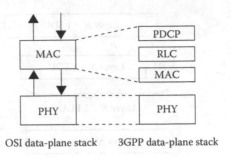

OSI data-plane stack   3GPP data-plane stack

Figure 2.7: NB-IoT data-plane protocol stack.

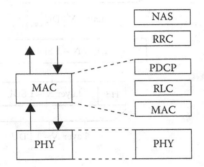

OSI control-plane stack   3GPP control-plane stack

Figure 2.8: NB-IoT control-plane protocol stack.

also explained. The book explains these sublayers that are introduced for 5G LTE NB-IoT network. The NAS layer is a signalling layer that evolved and exists with other 3GPP protocol stack such as UMTS™.

Figure 2.9 illustrates how a packet traverses the network layer (IP layer) and through the different NB-IoT data-plane protocol stack; PDCP, RLC, MAC, and PHY sublayers. Detailed explanation of each of these sublayers are to follow in the following chapters.

LTE has evolved as an enhancement to its predecessor system known as UMTS™. Enhanced UMTS™ Terrestrial Radio Access and Network (E-UTRA and E-UTRAN) are the official name used by 3GPP for LTE UE and its core network, respectively. E-UTRAN consists of an eNodeB that acts as a central controller (e.g., base station) connected to a large number of NB-IoT devices. Different eNodeBs are connected to each other and to the core network through protocols such as S1 and X2 protocols. The term E-UTRAN refers to the network side (eNodeB and core network), while the term E-UTRA refers to the UE side.

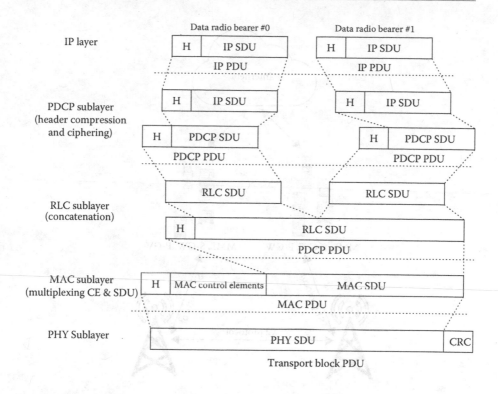

Figure 2.9: Packet traversal through NB-IoT protocol stack.

Figure 2.10 illustrates this architecture. Each eNodeB is responsible for providing radio covering to a geographical area, and all NB-IoT devices in this area can be connected to this eNodeB. A single or multiple eNodeBs belong to a mobile operator (e.g., AT&T, T-Mobile). All NB-IoT devices within the service area of a mobile operator are equipped and provided with an USIM card to enable their services on the mobile operator network.

The eNodeBs are interconnected with each other by means of the X2 protocol and interface. The eNodeBs are also connected by means of the S1 interface to the EPC (Evolved Packet Core)[4] which is the core network. More specifically, eNodeB is connected to the MME (Mobility Management Entity) by means of the S1-MME interface and to the Serving Gateway (S-GW) by means of the S1-U interface. The S1-MME interface carries control-plane messages and signalling, while the S1-U interface carries the data-plane messages.

---

[4]The term EPC or Core Network are used interchangeably.

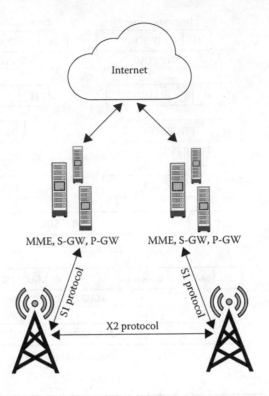

Figure 2.10: LTE NB-IoT network architecture.

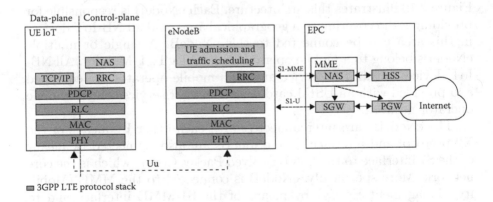

Figure 2.11: 3GPP LTE NB-IoT protocol stack for both UE and eNodeB.

Figure 2.11 illustrates the overall 3GPP protocol stack at the NB-IoT UE, eNodeB, and core network (EPC). The LTE core network, known as Evolved Packet Core (EPC), has two interfaces with the eNodeB; S1-MME protocol carries all signalling message, and S1-U carries all user or data messages. Data-plane traffic flows from the UE to eNodeB through the S1-U interface to the S-GW, Packet Gateway (P-GW), and finally to the Internet. Control-plane traffic flows from the UE to the eNodeB through the S1-MME interface to the MME.

MME is a control-plane component as it contains the NAS which is an anchor point for signalling messages exchanges with the UE. The MME can be overwhelmed by a large number of communications from NB-IoT devices since the number of NB-IoT devices within an MME region can be hundreds of thousands of devices. To handle such a large number of NB-IoT devices, there can be multiple MMEs communicating with the same eNodeB and performing load-balancing among themselves. MME communicates also with S-GW and P-GW. MME main functionalities are:

- NAS signalling (e.g., attach and tracking area update procedures, bearer establishment, and release).

- Authorization and authentication.

- Selection of S-GW and P-GW

- Lawful interception of signalling messages or data-plane message piggybacked with signalling messages.

Serving Gateway (S-GW) is the first component in the EPC that receives the data-plane packets from the UE through the S1-U interface. If data-plane packets of UE are piggybacked with NAS signalling messages, then those packets do not go through S-GW. S-GW main functionalities are as follows:

- Packet forwarding and routing to P-GW.

- Accounting for UE traffic.

- Local mobility anchor. If the UE moved to a different EPC, its traffic is routed through its home S-GW.

- Lawful interception of data-plane packets.

Packet Gateway (P-GW), which is a gateway to the Packet Data Network (PDN), is the second gateway in EPC. It acts as the access point for providing connectivity to the UE to the Internet, applications, and services. P-GW main functions are as follows:

- Support of IPv4, IPv6, DHCPv4, DHCPv6, and allocating an IP address to UE.

- Mapping of EPS bearer QoS parameters (QCI and ARP) to DiffServ Code point.

- Packet filtering and inspection.

- Data rate enforcement for a UE in both downlink and uplink.

- Accounting for UE traffic volume for both downlink and uplink.

- Lawful interception of data-plane packets.

HSS (Home Subscriber Server) is another EPC component used for storing and updating UE subscription information. HSS also stores UE information where different security keys for identity and traffic encryption are generated. HSS main functions are as follows:

- UE identification and addressing. It contains IMSI (International Mobile Subscriber Identity) or mobile telephone number.

- UE profile information. This includes UE-subscribed quality of service information (such as maximum allowed bit rate or allowed traffic class).

- Provide authentication between MME and UE.

- Provide the security keys used for ciphering and integrity-protecting signalling and data-plane messages exchanged between the UE and eNodeB.

In this book, we describe the 3GPP LTE NB-IoT protocol stack at the NB-IoT device side. It is worth noting that the same NB-IoT device stack exists and mirrored at the eNodeB. However, at the eNodeB side, there are multiple instances of the stack; one for each NB-IoT device.

2.8   NB-IoT Modes of Operation

The radio interface of the NB-IoT can support three modes of operation as illustrated in Figure 2.12. The following are the modes supported by an NB-IoT device:

- **Inband**: Utilizing the band of an LTE frequency. It utilizes resource blocks within an LTE carrier bandwidth where one physical resource block of LTE occupies 180 KHz of bandwidth.

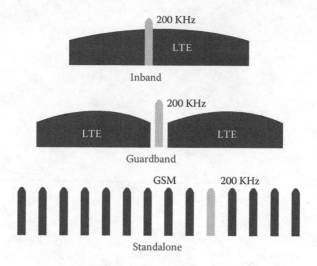

Figure 2.12: NB-IoT modes of operation.

- **Guardband:** Utilizing the band of an LTE frequency. It utilizes the unused (guard) resource blocks within an LTE carrier's guardband.

- **Standalone:** Utilizing a dedicated carrier other than LTE (e.g., GSM®). It occupies one GSM channel (200 KHz).

For the inband mode, NB-IoT signal occupies 180 KHz or one Physical Resource Block (PRB) within the LTE bandwidth. When the PRB is not used for NB-IoT, eNodeB can schedule it for other LTE traffic.

Chapter 3

Radio Resource Control Sublayer

3.1 Radio Resource Control Sublayer Capability

LTE™ NB-IoT UE contains much fewer features than the other types of UE (e.g., legacy LTE UE, relay node, sidelink UE, or high-power UE). This is to keep the UE complexity much lower, suitable for extremely low power consumption, low data speed, and lower cost. Table 3.1 summarizes those features that are supported and not supported in LTE NB-IoT UE [15].

3.2 Signalling and Data Radio Bearer

Signalling Radio Bearers (SRBs) are the Radio Bearers (RBs) that are used by UE for transmitting and receiving RRC messages with the eNodeB. For NB-IoT UE, only the following radio bearers are defined:

- **SRB0**: Used for carrying RRC signalling message during transmit and receive. It is used for the exchanging of the RRCConenctionRequest, RRCConnectionSetup, RRCConnectionResumeRequest, RRCConnectionReject, RRCConnectionReestablishmentRequest, and RRCConnectionReestablishment messages with eNodeB.

Table 3.1   Features Supported by an LTE NB-IoT UE

Features	NB-IoT UE	Full-Fledged UE
Number of master information block messages	1	1
Number of system information block Type 1 messages	1	1
Number of additional system information block messages	9	>20
Paging	Yes	Yes
Connection establishment	Yes	Yes
Security activation	Yes	Yes
Connection reconfiguration	Yes	Yes
Counter check	Yes	Yes
Connection re-establishment	Yes	Yes
Connection release	Yes	Yes
Inter-RAT mobility	No	Yes
Measurement	No	Yes
DL information transfer	Yes	Yes
UL information transfer	Yes	Yes
UE capability transfer	Yes	Yes
UE positioning	Yes	Yes
CSFB to 1x parameter transfer	No	Yes
UE Information	No	Yes
Logged measurement configuration	No	Yes
Release of logged measurement configuration	No	Yes
Measurements logging	No	Yes
In-device coexistence indication	No	Yes
UE assistance information	No	Yes
Mobility history information	No	Yes
RAN-assisted WLAN interworking	No	Yes
SCG failure information	No	Yes
LTE-WLAN aggregation	No	Yes
WLAN connection management	No	Yes
RAN-controlled LTE-WLAN interworking	No	Yes
LTE-WLAN aggregation with IPsec tunnel	No	Yes
MBMS	No	Yes
SC-PTM	Yes	Yes
Relay node procedures	No	Yes
Sidelink	No	Yes
Closed subscriber group	No	Yes
Carrier aggregation (CA)	No	Yes

(Continued)

Table 3.1 (*Continued*)   **Features Supported by an LTE NB-IoT UE**

Features	NB-IoT UE	Full-Fledged UE
Dual connectivity (DC)	No	Yes
Guaranteed bit rate (GBR)	No	Yes
Extended access barring	No	Yes
Self-configuration and self-optimization	No	Yes
Measurement logging	No	Yes
Public warning systems (CMAS and ETWS)	No	Yes
Real-time services (including emergency call)	No	Yes
Circuit-switched services and fallback	No	Yes

- **SRB1bis**: Established implicitly when establishing SRB1 after the UE receives the RRCConnectionSetup. SRB1bis is the same as SRB1 except that it bypasses the PDCP layer. SRB1bis is used as long as security is not activated. If security is activated, SRB1bis is not used but only SRB1 is used.

- **SRB1**: Used for RRC signalling message transfer after the security is activated. SRB0 is used before Access Stratum (AS) security is activated and only SRB1 is supported after AS security is activated.

- **DRB0** and **DRB1**: A maximum of two data radio bearers are used for exchanging data messages with the eNodeB. For a UE to support two DRBs, it must have its multi-DRB capability enabled; otherwise, UE supports only a single DRB.

SRB0 and SRB1bis both use Transparent Mode (TM) at RLC sublayer. Table 3.2 shows RLC and MAC sublayers configuration parameters for SRB1. Each parameter is explained in Tables 5.1 and 6.1.

Table 3.2   **RRC Default Configuration Parameters for SRB1**

Sublayer	Parameter	Value
RLC	t-PollRetransmit	25000 ms
	maxRetxThreshold	4
	enableStatusReportSN-Gap	Disabled
	logicalChannelIdentity	1
MAC	priority	1
		Highest priority
	logicalChannelSR-Prohibit	True

3.3   RRC Modes of Operation

RRC has only two states: either IDLE state or CONNECTED state as in Figure 3.1. They are literally called IDLE mode and CONNECTED mode since each mode has its own behavior and procedures. The initial mode of the UE is IDLE mode which is the mode when the UE is first powered on or upon insertion of an USIM. The UE toggles between these two modes. It moves from the IDLE to CONNECTED mode when a connection is established and moves back to IDLE mode when a connection is released as in Figure 3.1. A third mode can be used by UE, which is a power-saving mode where the UE is powered-off while it remains registered with the network.

In each of the two modes, the UE can perform any of the following functionalities:

IDLE Mode:

- Selection and (Re)selection of eNodeB.

- Acquire Master Information Block (MIB-NB) and System Information Blocks (SIBs).

- Monitors the logical Paging channel (PCCH) to detect incoming calls or system information change.

CONNECTED Mode:

- Transfer and exchange of UE unicast data with the eNodeB.

- Monitors Narrowband Physical Downlink Control Channel (NPDCCH) to detect if any resource is assigned to the UE for transmission or reception of control and data messages.

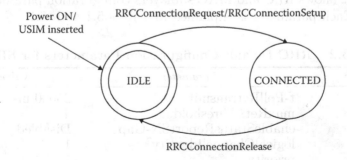

Figure 3.1: RRC modes of operation.

3.4 eNodeB Identities

The UE always keeps the following eNodeB identities to be used by the RRC and MAC sublayers:

- **RA-RNTI**: Identification used to scramble the NPDCCH during the random access procedure. RA-RNTI can be addressed to multiple UEs (i.e., multiple UEs can decode NPDCCH scrambled by the same RA-RNTI). RA-RNTI is 16 bits in length and its value can range from 1 to 960.

- **Temporary C-RNTI**: Identification used during the random access procedure. Temporary C-RNTI is 16 bits in length and its value can range from 1 to 65523.

- **C-RNTI**: Unique identification used for identifying UL and DL unicast transmissions. The eNodeB assigns different C-RNTI values to different UEs. C-RNTI is 16 bits in length and its value can range from 1 to 65523.

- **P-RNTI**: Identification used to scramble the NPDCCH when a Paging message or Direct Indication message are carried on NPDSCH (PCH). It is common for multiple UEs. It is 16 bits in length and its value is fixed to 65534.

- **SI-RNTI**: Identification used for broadcast of system information. It is 16 bits in length and its value is fixed to 65529.

- **SC-RNTI**: Identifies transmissions of the single-cell MCCH control information using SC-MCCH.

- **G-RNTI**: Identifies transmissions of a group MTCH information using SC-MTCH.

- **ResumeID**: 40-bit unique UE identification used for the RRC connection resume procedure.

- **IMSI**: 6–21-digit unique UE identification.

- **S-TMSI**: 40-bit unique identification.

The different RNTIs, used with NPDCCH, are explained in Section 7.10.9.

3.5 RRC PDU Format

RRC is a control-plane sublayer and it exchanges its messages with eNodeB in a form of a PDU without a header as in the format shown

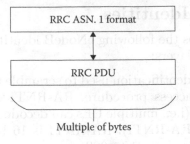

Figure 3.2: RRC PDU format.

in Figure 3.2. RRC PDU payload contains a signalling message[1] as will be explained in Section 3.7. RRC message is expressed in ITU ASN.1 format [16]. It uses Packed Encoding Rule (PER), which is a compact form of converting the message from human-readable format to binary format. PER is used by 3GPP™ as it results in a small binary PDU size which is suitable for a wireless link with limited bandwidth. While PER does not align the individual parameters (fields) of the RRC message to octet boundary, it aligns the final RRC PDU to an octet boundary.

RRC PDU does not have a header nor a trailer. After the RRC constructs or receives a message exchanged with eNodeB, it either encodes or decodes it, respectively, using ASN.1 PER format. RRC PDU is exchanged with PDCP sublayer if security is activated. If security is not activated, RRC PDU bypasses the PDCP sublayer and is exchanged with the RLC sublayer directly.

3.6   UE Behavior in IDLE Mode

3.6.1   *PLMN selection*

The NAS can maintain a list of PLMNs in priority order. PLMN represents a mobile operator in a geographical location. One of the PLMNs is selected either manually by the user or automatically to be the target PLMN. The RRC is then requested to search for a cell that belongs to this PLMN and retrieves all system information from that cell. If the cell is found to be suitable, the UE selects this cell, continuously monitors it, and acquires its MIB and SIB1. The cell is then identified as the serving cell and the UE is said to be camping on this cell. The UE needs to camp on a cell that belongs to the target PLMN. PLMN identities are broadcasted during SIB1 and the UE has to tune to the

---

[1]Message and PDU are the same data unit and can be used interchangeably.

broadcast channel and acquire MIB and SIB1 of the cell in order to identify PLMN of each detected cell. A single PLMN can have one or more cells and each cell can belong to one or more PLMN. The UE can scan all RF channels in the bands supported by the UE to find a cell that belongs to the target PLMN or select any cell if no PLMN is targeted. Upon scanning all RF channels, any detected PLMN is reported to the NAS provided that its received RSRP value at the UE exceeds −110 dBm [17, 18]. Those cells that have their received RSRP at the UE greater than −110 dBm are marked as high-quality cells. Cells that are less than −110 dBm can also be reported to the NAS but not marked as high-quality cells. Among those PLMN cells reported to the NAS, whether high quality or not, the NAS can select a target PLMN either automatically or manually. The RRC is then requested to find, select, and camp on a cell that belongs to the target PLMN [17].

3.6.2 Cell selection

Cell Selection refers to the procedure the UE performs when selecting a cell for the first time. That is, when the UE has not detected any cell yet, it does a cell selection to select a cell that belongs to the target PLMN to camp on. For cell selection, the UE needs to tune to each broadcast channel frequency, acquire MIB and SIB1, identify that the cell belongs to the target PLMN if specified, and if the cell meets the cell selection criteria, the UE selects the cell and camps on it. If the UE camps on a cell, this cell becomes the serving cell of the UE. If multiple cells exist that the UE can camp on, the UE typically chooses the strongest cell to camp on.

3.6.3 Cell reselection

Cell Reselection is different from cell selection as the latter is done during the first time or after the UE is powered on so that the UE can camp on a serving cell. After the first cell selection and camping on a cell, the UE periodically do a cell reselection to find if there is any other cell that can be stronger than the current serving cell. If a stronger cell is found, then the new cell is selected and becomes the new serving cell.

3.6.4 Suitable cell

In the IDLE mode, a UE starts searching for a suitable cell and camps on it. A suitable cell is a cell that fulfills the following conditions:

- Part of the selected PLMN.

- Is not a barred cell.

- Part of at least one Tracking Area (TA) as advertised in SIB1-NB and not among the forbidden tracking areas. A tracking area is a group of eNodeBs where a UE in IDLE mode do not perform Tracking Area Update (TAU) procedure.

- The S criteria is fulfilled.

- The cell is a high-quality cell.

PLMN selection is done either manually as the user selects a desired PLMN, or automatically through NAS. The NAS can request to register for a specific PLMN, if the PLMN is stored on the USIM. If a PLMN is not selected by any means, the UE can select any PLMN and find a suitable cell for it. The UE can detect whether a cell is barred or not barred through the information received in SIB1-NB. Finally, the S criteria for a cell is that the received RSRP at the UE satisfy the following condition:

$$S_{rxlev} > 0 \quad \text{AND} \quad S_{qual} > 0, \tag{3.1}$$

where:

$$S_{rxlev} = Q_{rxlevmeas} - Q_{rxlevmin},$$
$$S_{qual} = Q_{qualmeas} - Q_{qualmin}.$$

S_{rxlev} and S_{qual} are the cell Rx level and quality values, respectively, in dB and the other parameters are received by the UE through the broadcasted SIBs as defined as in Table 3.3.

At the UE, if any cell satisfies Equation (3.1), then it is a candidate for being a suitable cell. The behavior of the UE in IDLE mode is best illustrated as in Figure 3.3.

Table 3.3 *S* **Criteria Parameters**

Parameter	RRC	Meaning
$Q_{rxlevmeas}$	-	Measured cell Rx level value (RSRP) by UE
$Q_{qualmeas}$	-	Measured cell quality value (RSRQ) by UE
$Q_{rxlevmin}$	SIB1-NB, SIB3-NB, SIB5-NB	Minimum required Rx level in the cell (dBm). Other values exist in SIB3-NB and SIB5-NB for intra- and inter-frequency Cell (Re)selection evaluation
$Q_{qualmin}$	SIB1-NB, SIB3-NB, SIB5-NB	Minimum required quality level in the cell (dB). Other values exist in SIB3-NB and SIB5-NB for intra- and inter-frequency Cell (Re)selection evaluation

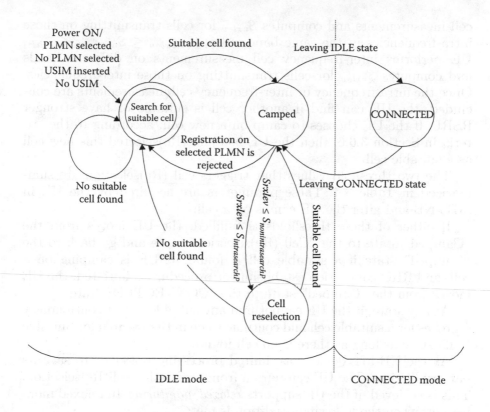

Figure 3.3: UE behavior in IDLE mode.

When the UE is first powered on, a PLMN is selected either manually or automatically, or if an USIM is inserted, the stored PLMN is selected. The UE starts searching for a suitable cell for the target PLMN by scanning the RF frequencies of the target PLMN on all bands supported by the UE. If no PLMN is targeted, then the UE scans all RF frequencies until a suitable cell is found. During scanning of the RF frequencies, the UE acquires MIB-NB and SIB1-NB of the corresponding cell so that it can determine whether the cell is barred or not and computes the S criteria. When a cell is found to be suitable, the UE moves to the "Camped" state and camps on that cell. The cell become the serving cell of the UE.

3.6.5 Triggers for cell reselection for inter-frequency and intra-frequency cells

Periodically and during the time UE is camping on a cell, the UE checks S_{rxlev} value and if $S_{rxlev} \leq S_{IntraSearchP}$, UE performs intra-frequency cell measurements and computes S_{rxlev} for cells transmitting on those intra-frequencies in the same band. Also, if $S_{rxlev} \leq S_{nonintraSearchP}$, UE performs inter-frequency cell measurements on different bands and computes S_{rxlev} for cells transmitting on those inter-frequencies. Once the intra-frequency or inter-frequency cell measurements are concluded, the UE can find if another cell is suitable or have stronger RSRP. If the UE chooses to camp on a new cell, according to the criteria in Section 3.6.6, then the UE has just (Re)selected this new cell as a suitable cell.

The two threshold values that triggers cell (Re)selection are summarized in Table 3.4. These parameters are acquired by the UE in SIB3-NB and after the UE camps on a cell.

If either of these thresholds is fulfilled, the UE moves from the "Camped" state to the "Cell (Re)selection" state and go back to the "Camped" state if a suitable cell is found. If UE is camping on a cell, and RRC connection establishment procedure is initiated, the UE moves from the "Camped" state to the "CONNECTED" state.

At any state, if the UE cannot find any suitable cell, it continuously searches for a suitable cell and continue to be in the "Search for Suitable Cell" state as long as there is no cell found.

As the NB-IoT device has limited processing power or to save its battery lifetime, The UE can refrain from performing cell(Re)selection. This is achieved, if the UE supports *relaxed monitoring*. In relaxed monitoring, when the following condition is met:

$$S_{rxlevRef} - S_{rxlev} < S_{SearchDeltaP},$$

for the last 24 hours, the UE does not need to check the triggers of performing cell (Re)selection and hence refrain from doing cell (Re)selection. $S_{rxlevRef}$ is the value of S_{rxlev} for the serving cell when performing the last cell selection or reselection. $S_{SearchDeltaP}$ as in Table 3.4.

Table 3.4   Cell (Re)selection Triggers

Parameter	RRC	Meaning
$S_{NonIntraSearch}$	SIB3-NB	Threshold Rx level value (RSRP) that triggers cell (Re)selection for Inter-frequency
$S_{IntraSearchP}$	SIB3-NB	Threshold Rx level value (RSRP) that triggers cell (Re)selection for Intra-frequency
$S_{SearchDeltaP}$	SIB3-NB	Threshold Rx level value (RSRP) that prevents cell (Re)selection

3.6.6 Cell reselection for inter-frequency and intra-frequency cells

The criteria used for (Re)selecting a suitable cell is different than the criteria used for selecting a cell. Cell reselection occurs when a UE had already selected a cell and camped on that cell. If the threshold triggers in Table 3.4 is met, the UE starts to perform a cell (Re)selection. All suitable cells detected by the UE are ordered in a decreasing order of their ranking values where the ranking values of the serving cell and neighbor cells are calculated according to the following:

$$R_s = Q_{meas,s} + Q_{Hyst},$$
$$R_n = Q_{meas,n} - Q_{offset}. \qquad (3.2)$$

The evaluation of Equation (3.2) only takes place for a cell that meets the S criteria. If a cell with highest ranking values is higher in ranking than the current serving cell for a time interval $T_{reselection}$ and the UE has camped on the current cell for more than one second, the new highest ranking cell is selected to be the new serving cell. The parameters used in the above equation are explained in Table 3.5.

3.7 RRC Procedures and Behavior in CONNECTED Mode

3.7.1 Master information block (MIB-NB)

The MasterInformationBlock-NB (MIB-NB) and SystemInformation-BlockType1-NB (SIB1-NB) are the first messages to be acquired by a UE when it is powered on or when an USIM is inserted. These two messages contain important information about the cell and eNodeB to be accessed. An eNodeB transmits these messages in a repetitive

Table 3.5   Cell (Re)selection Parameters

Parameter	RRC	Meaning
Q_{meas}	-	RSRP measurement quantity by UE that is used in cell reselections for serving cell ($Q_{meas,s}$) and neighbor cells ($Q_{meas,n}$)
Q_{Hyst}	SIB3-NB	A hysteresis value used to prevent ping-pong effect of selecting and reselecting the same cell
Q_{offset}	SIB5-NB	A resource or frequency offset to apply (dB)
$T_{reselection}$	SIB3-NB, SIB5-NB	Reselection timer (in seconds) for either Intra- or Inter-frequency cells, respectively

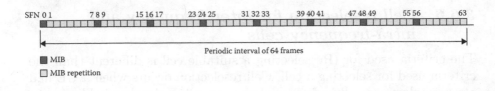

Figure 3.4: MIB-NB scheduling.

way to increase the probability of a UE being able to acquire them reliably. MIB-NB is scheduled for transmission by an eNodeB periodically. Each 64 radio frames (640 ms), eNodeB transmits a MIB-NB message. The first radio frame in this 64 radio frames is scheduled to carry a partial MIB-NB message while subsequent radio frames contain repetitions of the partial MIB-NB message. The first transmission of MIB-NB occurs in the first subframe where the system frame number is SFN mod 64 = 0.

The MIB-NB message is divided into 8 equal blocks by the physical sublayer. The first block is transmitted on the first subframe in a frame and repeated in the first subframe in the next seven consecutive frames. That is, each block is transmitted and repeated in 8 frames (80 ms). Figure 3.4 illustrates the MIB-NB scheduling. It shows the first transmission of the MIB-NB and the repetitions in each frame.

The parameters of the MIB-NB message are shown in Table 3.6. systemFrameNumber-MSB contains the 4 MSB of the SFN. The 6 LSB are retrieved implicitly by the UE when decoding NPBCH. The total size of the MIB-NB is 34 bits.

Time is divided into a number of hyper system frames where each hyper system frame consists of 1024 system frames. Both Hyper System Frame Number (H-SFN) and System Frame Number (SFN) are 10 bits that range from 0 to 1023. H-SFN has a cycle of 10.24 seconds since each system frame is 10 ms.

3.7.2 *System information block type 1 (SIB1-NB)*

System Information Block Type 1 (SIB1-NB) contains information whether the UE is allowed to access a cell or not. In addition, it includes information and parameters about the time scheduling of other system information blocks. SIB1-NB follows a similar scheduling pattern to MIB-NB but with different periodicity. For each 256 radio frames (2560 ms), eNodeB transmits a SIB1-NB message. The 256 frames are grouped into 16 groups of frames where each group consists of 16 radio frames. One of the frame, called the starting frame, in the 16 radio frames contains the first SIB1-NB transmission, while remaining frames

Table 3.6   **MIB-NB Parameters**

Parameter	Size (Bits)	Meaning
systemFrame-Number-MSB	4	4 most-significant bits of the 10 bits representing SFN
hyperSFN-LSB	2	2 least significant bits of hyper SFN. The remaining bits are present in SIB1-NB
systemInfoValueTag	5	A value that is incremented if any of the SIB contents have changed
schedulingInfoSIB1	4	An index value that is used to determine how SIB1-NB is scheduled
ab-Enabled	1	If true, indicates that access barring to this eNodeB is enabled
operationModeInfo	7	Determine whether the cell operates in one of the following mode: -**Inband (SamePCI)**: NB-IoT and LTE cell share the same physical cell ID -**Inband (DifferentPCI)**: NB-IoT and LTE cell have different physical cell ID -**Guardband**: a guardband deployment -**Standalone**: a standalone deployment
Spare	11	For future extension

contain other transmission of SIB1-NB. The starting frame and number of repetition within those 16 frames are configured by eNodeB. SIB1-NB transmission always occurs in subframe #4 in a frame that contains SIB1-NB transmission.

Figure 3.5 shows an example of the SIB1-NB scheduling. In this figure, SIB1-NB periodicity is 256 radio frames. Each of the 16 groups consists of 16 radio frames, the first SIB1-NB transmission occurs at the first frame in the group. SIB1-NB is only transmitted in subframe #4 of every other frame in the group. SIB1-NB repetitions and the number of repetitions within the 256 frames are equally spaced within the 256 frames.

The starting frame and number of repetition for the SIB1-NB are determined based on schedulingInfoSIB1 (as in Tables 3.7 and 3.8) and Physical cell ID. Figure 3.5 illustrates an example of SIB1-NB scheduling when schedulingInfoSIB1 is 2 (16 repetitions) and starting

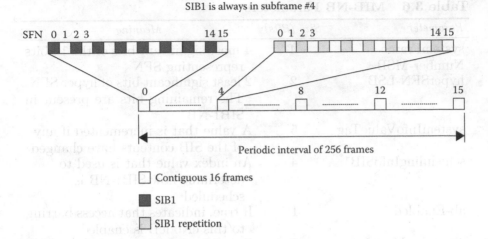

Figure 3.5: SIB1-NB scheduling.

Table 3.7 SchedulingInfoSIB1 Values

SchedulingInfoSIB1 Values	Number of SIB1 Repetitions
0	4
1	8
2	16
3	4
4	8
5	16
6	4
7	8
8	16
9	4
10	8
11	16

radio frame is zero (PHY_{ID}^{CELL} mod 2 = 0 as in Table 3.8). SIB1-NB repetitions are repeated in each group within the 256 frames. Table 3.9 shows the content of an SIB1-NB message.

3.7.3 Other system information block

NB-IoT UE may also need to acquire other systems information blocks. These are shown in Table 3.10 with the purpose of acquiring them. For scheduling these SIBs, time is divided into an equal number of

Table 3.8   Starting Radio Frame for SIB1-NB

Number of SIB1 Repetitions	PHY_{ID}^{CELL}	Starting Radio Frame Number for SIB1
4	PHY_{ID}^{CELL} mod $4 = 0$	0
	PHY_{ID}^{CELL} mod $4 = 1$	16
	PHY_{ID}^{CELL} mod $4 = 2$	32
	PHY_{ID}^{CELL} mod $4 = 3$	48
8	PHY_{ID}^{CELL} mod $2 = 0$	0
	PHY_{ID}^{CELL} mod $2 = 1$	16
16	PHY_{ID}^{CELL} mod $2 = 0$	0
	PHY_{ID}^{CELL} mod $2 = 1$	1

Table 3.9   SIB1-NB Parameters

Parameter	Size (Bits)	Meaning
hyperSFN-MSB	8	8 MSB of hyper-SFN. The 2 LSB are indicated in MIB-NB (Table 3.6). This constructs a 10 bit Hyper SFN. Hyper-SFN is incremented by one when the SFN wraps around
plmn-IdentityList	List	A list of PLMN ID where this cell belongs to. PLMN ID consist of a 3-digit MCC and 2 or 3-digit MNC
trackingAreaCode	16	A Tracking Area Code (TAC) that is common to all PLMNs in the list
cellIdentity	28	A cell ID that is unique within a PLMN
cellBarred	1	Whether this cell is barred or not
si-WindowLength	3	Size of SI Window in milliseconds where only one SI is scheduled within the Window. Values as in Table 3.11
si-TB	3	Indicates transport block size, in bits, for each SI message. Values as in Table 3.11
schedulingInfoList	List	A list that contains scheduling information about SIB2-NB to SIB22-NB (some of the scheduling information is shown in Table 3.11)
systemInfoValue-TagList	List	A list of SystemInfoValueTagSI for each SIB that indicates if the corresponding SIB has its content changed by eNodeB

Table 3.10 Optional System Information Blocks

System Information Block	Purpose
SystemInformationBlockType2-NB (SIB2-NB)	Contains radio resource configuration for PDCP, RLC, MAC, and PHY sublayers that are common for all UEs. It also contains information about the network support for CIoT optimization, random access and DRX power saving parameters
SystemInformationBlockType3-NB (SIB3-NB)	Contains common cell (Re)selection information for intra- and inter-frequency cell (Re)selection other than for neighbouring cells
SystemInformationBlockType4-NB (SIB4-NB)	Contains neighbouring cell related information relevant only for intra-frequency cell (Re)selection
SystemInformationBlockType5-NB (SIB5-NB)	Contains neighbouring cell related information relevant only for inter-frequency cell (Re)selection
SystemInformationBlockType14-NB (SIB14-NB)	Contains Access Barring parameters
SystemInformationBlockType15-NB (SIB15-NB)	Used if UE supports MBMS. This SIB indicates MBMS Service Area Identities (SAI) of the current and neighbouring carrier frequencies
SystemInformationBlockType16-NB (SIB16-NB)	Contains information related to GPS time and Coordinated Universal Time (UTC)
SystemInformationBlockType20-NB (SIB20-NB)	Used if UE supports MBMS. It contains information to acquire SC-MCCH
SystemInformationBlockType22-NB (SIB22-NB)	Used if UE supports Paging and RACH on non-anchor carriers

frames called SI Window length (W) (si-WindowLength in Table 3.9). Each Window length contains only one System Information (SI) RRC message where each SI message can contain one or more SIBs. That is, SI messages are not overlapped in time and at most one SI message is transmitted within each SI Window length.

Table 3.11 shows the scheduling information used to schedule SIB2-NB to SIB16-NB transmission. These are the configuration parameters

Table 3.11   SIB2-NB to SIB22-NB Scheduling Information

Parameter	Meaning	Possible Values	Example Values
si-WindowLength	Equal interval of times (ms) where each interval contains at most one SIB. A single value for all SIBs (si-WindowLength in Table 3.9)	160, 320, 480, 640, 960, 1280, 1600	160
n	The order of the SIB as broadcasted in SIB2	[1, 8]	1 for SIB2-NB 1 for SIB3-NB 2 for SIB4-NB 3 for SIB5-NB 4 for SIB14-NB 5 for SIB16-NB SIB15-NB, SIB20-NB, and SIB22-NB are not transmitted
si-RadioFrame Offset	Starting frame offset within each Window length	[1, 15] if absent means zero	1

For each SIB:

Parameter	Meaning	Possible Values	Example Values
Periodicity	Periodicity of a SIB in RF	64, 128, 256, 512, 1024, 2048, 4096	128
Repetition Pattern	How a SIB is repeated within each SI Window length	every2ndRF, every4thRF, every8thRF, every16thRF	Every 2nd frame
si-TB	Transport block size	b56, b120, b208, b256, b328, b440, b552, b680	56 bits is the transport block size for a SIB

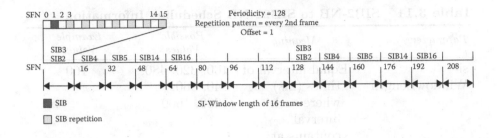

Figure 3.6: SIB2-NB to SIB16-NB scheduling per scheduling information in Table 3.11.

that are used to determine how SIB2-NB to SIB16-NB are scheduled by eNodeB and hence aids the UE in acquiring them.

Figure 3.6 illustrates an example of how SIB2-NB to SIB16-NB are scheduled given the configuration parameters in Table 3.11. SIB2-NB has a special case where it is always co-located in the first entry of the list, schedulingInfoList, as in Table 3.9. Hence, SIB2-NB and SIB3-NB are co-located together in the same SI message and same window, si-WindowLength, and this indicates n equals to 1 for both SIB2-NB and SIB3-NB.

Note that SIB15-NB, SIB20-NB, and SIB22-NB are not transmitted by eNodeB. As the Window length is 16, each 16 frames, at most one SI message is scheduled. The first SI message transmitted contains both SIB2-NB and SIB3-NB. In each 16 frames, the offset of the frame where the SI message is scheduled is in the second frame (frame number 1). The same SI message is repeated every second frame. Periodicity of all SI messages is 128 frames which means any SI message is repeated every 128 frames. si-TB determines the transport block size received at the physical sublayer for each SI message. In this example, si-Tb is the same for all scheduled SI messages and is 56 bits.

3.7.4 System information modification period

eNodeB can change the content of MIB-NB or one or more SIB-NB messages. This can be indicated to the UE using direct indication message or in a Paging message. Change in SIBs occurs only at a specified interval known as a modification period as shown in Figure 3.7. The modification period occurs every 4096 frames. If the SIB content is to be changed, this is first indicated to UE in a modification period n, the next modification period, $n + 1$, contains the new updated SIB. The modification period boundary is defined by SFN values for which (H-SFN * 1024 + SFN) mod 4096 = 0. That is, the modification period boundary

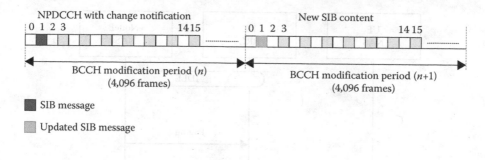

Figure 3.7: SIB-NB modification period.

is each 40.96 seconds. The MIB-NB contains the 2 LSB of Hyper System Frame Number (H-SFN) and SIB1-NB contains the 8 MSB which if combined together and with the SFN, the UE can infer the modification period boundary.

In IDLE mode, UE can acquire and re-acquire system information when they are detected to have changed; otherwise, if the UE is in CONNECTED mode, UE does not need to re-acquire the SIBs. However, if essential information in SIBs have changed while in CONNECTED mode, eNodeB can trigger RRC connection release procedure.

If eNodeB is to change MIB-NB or SIB-NB information, it indicates such a change in a Paging or direct indication message.

3.7.5 Paging

The purpose of this procedure is to inform the UE about incoming call, outgoing call, or about a change in any of the system information in MIB-NB, SIB1-NB, or other SIBs. This procedure applies only when the UE is in IDLE mode. This procedure is not invoked while the UE is in CONNECTED mode. If the Paging message is for incoming or outgoing call, the upper layer of the UE (NAS) is informed and can initiate a connection establishment procedure. Paging procedure is shown in Figure 3.8.

If the NB-IoT UE is in CONNECTED mode, it does not require to detect SIB changes. eNodeB can release the RRC connection and let the UE move to IDLE mode to acquire changed SIB(s) [2].

UE in IDLE mode receives Paging message on the anchor or non-anchor carrier. Paging message is detected by the UE by monitoring the NPDCCH scrambled with P-RNTI. Table 3.12 shows the content of the Paging message. Paging can be for a single UE or multiple UEs. UE is identified by its ID (S-TMSI or IMSI) and all UEs being Paged are included in the pagingRecordList. The Paging message can be also

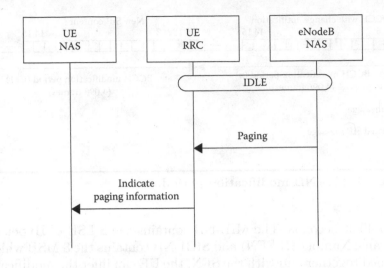

Figure 3.8: Paging procedure.

Table 3.12   Paging Message

Parameter	Size (Bits)	Meaning
pagingRecordList	List	A list of IMSI or S-TMSI of the UEs being Paged
systemInfoModification	1	If presents, indicates a MIB or other SIBs modification other than for SIB14 and SIB16

an indication of a change in system information (MIB-NB and all SIBs except SIB14-NB and SIB16-NB) in which case the systemInfoModification parameter is included.

3.7.6   RRC direct indication information

Direct indication information is another form of RRC Paging message transmitted on the NPDCCH to indicate a change in system information (SIBs). It has only two bits and if any one of them is set to 1, it indicates a change in system information and the UE can re-acquire SIBs again.

3.7.7   RRC connection establishment

The purpose of this procedure is to establish a new connection to the eNodeB. That is, this procedure also moves the UE from IDLE mode to the CONNECTED mode. After completion of this procedure, UE establishes two SRBs: SRB1 and SRB1bis.

Figure 3.9 illustrates this procedure. Connection is established upon a request from NAS and once connection establishment is completed, NAS is informed. When the UE sends the RRCConnectionRequest message, it indicates the reason for establishing this connection which can be for mobile-originated signalling or data, mobile-terminated signalling or data, mobile-originated exceptional data, or delay tolerance access. RRCConnectionRequest message content is summarized in Table 3.13.

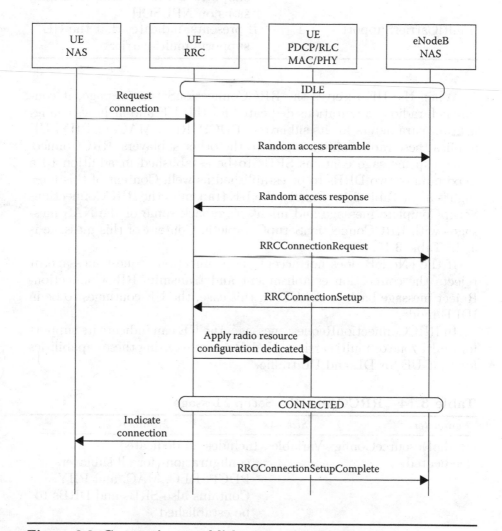

Figure 3.9: Connection establishment procedure.

Table 3.13   RRCConnectionRequest Message

Parameter	Size (Bits)	Meaning
ue-Identity	40	S-TMSI or 40 bit random value identification of the UE
establishmentCause	3	Indicates type of access (Mobile terminated access, Mobile originating signalling, or data, or Exceptional data, Delay tolerant access)
multiToneSupport	1	If presents, indicates that the UE supports UL multi-tone transmissions on NPUSCH
multiCarrierSupport	1	If presents, indicates that the UE supports multi-carrier

When the UE receives the RRCConnectionSetup message, it contains all radio configurations dedicated for this UE and includes configuration parameters for its sublayers: PDCP, RLC, MAC, or PHY. UE applies these radio configurations to the other sublayers. RRCConnectionSetup message contains SRBs to be established in addition to a maximum of two DRBs to be established as well. Content of this message is as in Table 3.14. Finally, the UE transmits the RRCConnectionSetupComplete message and might piggyback some of the NAS messages with RRCConnectionSetupComplete. Content of this message is as in Table 3.15.

If the eNodeB does not accept the connection request message, it rejects the connection establishment and transmits RRCConnectionReject message back to the UE. In this case, the UE continues to be in IDLE mode.

In RRCConnectionRequest message, the UE can indicate its support for multi-tone or multi carrier. eNodeB can start using these capabilities for that UE for DL and UL traffic.

Table 3.14   RRCConnectionSetup Message

Parameter	Size	Meaning
RadioResourceConfig Dedicated	Variable	Includes all dedicated configurations for all sublayers; PDCP, RLC, MAC, and PHY. Contains also SRBs and DRBs to be established

Table 3.15 RRCConnectionSetupComplete Message

Parameter	Size (Bits)	Meaning
s-TMSI	40	Assigned S-TMSI of the UE
dedicatedInfoNAS	Variable	Carries NAS information piggybacked with this RRC message
up-CIoT-EPS-Optimization	1	If presents, indicates if the UE supports User plan CIoT Optimization or S1-U data transfer

3.7.8 Initial security activation

When the UE is in CONNECTED mode, the eNodeB can establish the security for the SRB1 and any DRBs. Security activation means that the UE applies integrity and ciphering algorithms to incoming or outgoing signalling or data messages.

Figure 3.10 illustrates the security activation procedure. After the connection establishment procedure is completed, and when the UE receives the SecurityModeCommand, it derives a number of keys that are used by the integrity and ciphering algorithms at the PDCP sublayer. After receiving the SecurityModeCommand, the UE derives the key, K_{eNB}. From the K_{eNB}, UE derives the integrity key, K_{RRCint}, and uses this key to verify the integrity of the received SecurityModeCommand. If SecurityModeCommand message passes the integrity check, the UE derives K_{RRCenc} and K_{UPenc} which are the keys used for ciphering RRC messages and data-plane traffic, respectively. SecurityModeCommand message is shown in Table 3.16.

By having these keys, the UE starts integrity-protecting and ciphering all signalling and data message including the outgoing SecurityModeComplete message. By completing this procedure, the security is considered as activated at the UE and the UE can start exchanging control and data messages securely with the eNodeB. In addition, when this procedure completes, the SRB1bis is no longer used and UE starts to use SRB1.

Table 3.16 SecurityModeCommand Message

Parameter	Size (Bits)	Meaning
ciphering Algorithm	4	Indicates cipher algorithm to be used for ciphering signalling and data RB. Possible values are as in Table 3.18
integrityProt Algorithm	4	Integrity algorithm to be used for protecting signalling RB. Possible values are as in Table 3.17

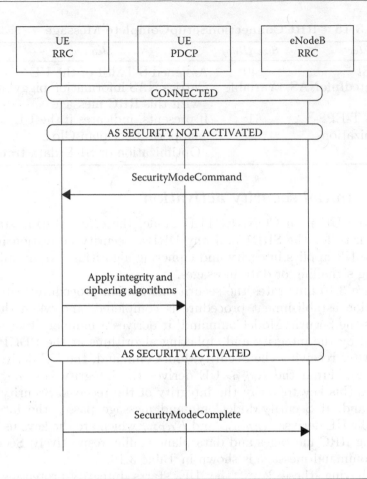

Figure 3.10: Security activation procedure.

Both the EPS Integrity Algorithms (EIA) and the EPS Encryption Algorithms (EEA) supported by the UE are illustrated in Tables 3.17 and 3.18, respectively.

Table 3.17   EPS Integrity Algorithm (EIA)

Parameter	Meaning
EIA0	Null ciphering algorithm
128-EIA1	SNOW 3G based algorithm
128-EIA2	AES based algorithm
128-EIA3	ZUC based algorithm

Table 3.18 **EPS Encryption Algorithm (EEA)**

Parameter	Meaning
EEA0	Null ciphering algorithm
128-EEA1	SNOW 3G based algorithm
128-EEA2	AES based algorithm
128-EEA3	ZUC based algorithm

3.7.9 RRC connection resume

The purpose of this procedure is to resume a connection to the eNodeB that has been suspended (through the RRC connection release procedure with releaseCause set to rrc-Suspend). If a UE connection is suspended, it moves from the CONNECTED mode to IDLE mode. If the connection latter is resumed, the UE moves back from IDLE mode to the CONNECTED mode. After completion of this procedure, UE establishes SRB1 and any DRB. If a connection is suspended, both the UE and eNodeB saves the AS context of the UE.

Figure 3.11 illustrates this procedure. Connection resumption is initiated upon a request from NAS and once connection resumption is completed, NAS is informed. When the UE sends the RRCConnectionResumeRequest message, it indicates the reason for resuming the connection which can be for mobile-originated signalling or data, mobile-terminated signalling or data, mobile originated exceptional data, or delay tolerance access. RRCConnectionResumeRequest message content is summarized in Table 3.19. The UE has its security activated previously before the connection is suspended and the RRCConnectionResumeRequest is sent with a message authentication code (MAC).

When the UE receives the RRConnectionResume message, it updates and restores the security context and updates the ciphering and integrity keys: K_{RRCint}, K_{RRCenc}, and K_{UPenc}. The RRConnectionResume message also contains dedicated radio configuration for this UE and includes configuration parameters for its sublayers: PDCP, RLC, MAC, or PHY. UE applies these radio configurations to other sublayers. UE resumes all DRBs. Content of this message is as in Table 3.20.

Finally, the UE transmits the RRConnectionResumeComplete message and it might piggyback some of the NAS messages with RRCConnectionResumeComplete. Content of this message is as in Table 3.21.

The UE can choose to resume the RRC connection to another new eNodeB other than the old eNodeB where the RRC connection was suspended before. Both new and old eNodeB communicate together

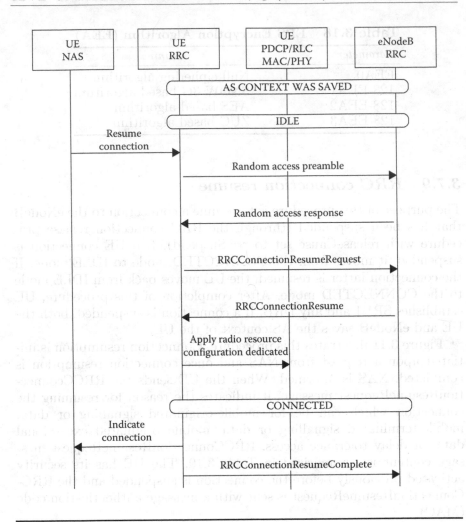

Figure 3.11: Connection resume procedure.

so that the new eNodeB can retrieve the UE-stored information and context from the old eNodeB.

3.7.10 RRC connection reconfiguration

When the UE is in CONNECTED mode and after the security has been activated, eNodeB might need to change the configuration parameters of PDCP, RLC, MAC, or PHY sublayers. In addition, eNodeB might need to establish or release any signalling or data radio bearer.

For such a purpose, eNodeB transmits an RRCConnectionReconfig-uration message to the UE. Figure 3.12 illustrates this procedure. If the

Table 3.19   RRCConnectionResumeRequest Message

Parameter	Size (Bits)	Meaning
resumeID	40	An ID to identify the AS context of the UE
resumeCause	3	Indicates type of access (Mobile terminated access, Mobile originating signalling, Data, Exception data, or Delay tolerant access)
shortResumeMAC-I	16	MAC-I used to identify and verify the UE

Table 3.20   RRCConnectionResume Message

Parameter	Size (Bits)	Meaning
RadioResourceConfig Dedicated	Variable	Includes all dedicated configurations for all sublayers: PDCP, RLC, MAC, and PHY. Contains also SRBs and DRBs to be resumed

Table 3.21   RRCConnectionResumeComplete Message

Parameter	Size (Bits)	Meaning
selectedPLMN-Identity	3	Indicates index of the PLMN selected by the UE from the plmn-IdentityList included in SIB1-NB
dedicatedInfoNAS	Variable	Carries NAS information piggybacked with this RRC message

UE receives an RRCConnectionReconfiguration message, it applies the new radio configurations to PDCP, RLC, MAC, or PHY sublayers including any establishment or release of any radio bearer. Also, any NAS message can be piggybacked with the RRCConnectionReconfiguration message.

Table 3.22 shows the content of the RRCConnectionReconfiguration Message.

3.7.11   RRC connection re-establishment

The purpose of this procedure is to resume connection to the eNodeB after an error condition that has happened and caused the UE to lose connection temporary with the eNodeB. During the error, the UE is not able to have a communication with the eNodeB. Error conditions include lost signal, weak signal, integrity-check failures, excessive

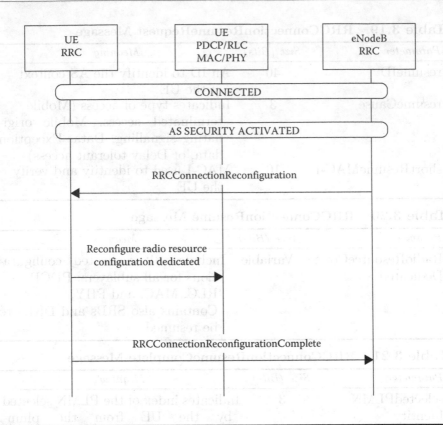

Figure 3.12: Connection reconfiguration procedure.

Table 3.22 RRCConnectionReconfiguration Message

Parameter	Size	Meaning
dedicatedInfoNAS	Variable	Carries NAS information piggy-backed with this RRC message
RadioResourceConfig Dedicated	Variable	Includes all dedicated configurations for all sublayers: PDCP, RLC, MAC, and PHY. Contains also SRBs and DRBs to be reconfigured

number of messages transmitted to the eNodeB without receiving any acknowledgment for them, or a radio link failure. The UE can also initiate this procedure if it receives radio configuration parameters that the UE cannot comply with. The reason for initiating this procedure is as in Table 3.23.

In addition, this procedure can be triggered by the UE, that supports Control place IoT EPS optimization, when the AS security is not

Table 3.23 **RRCConnectionReestablishmentRequest** Message

Parameter	Size (Bits)	Meaning
ReestablishmentCause	2	Indicates the failure cause that triggered the re-establishment procedure. Possible values are {reconfigurationFailure, otherFailure}
ue-Identity	S-TMSI	UE identity included to retrieve UE context and at the eNodeB to facilitate contention resolution by lower layers

activated to resume operation on SRB1bis and continue traffic transfer on this SRB1bis. A UE that has activated security uses this procedure to re-establish the SRB1.

If this procedure is initiated due to any error condition, the UE starts cell search and random access procedure at the MAC sublayer as if the UE is starting to move from IDLE mode to the CONNECTED mode for the first time. This procedure is used by the UE if it is in CONNECTED mode and security has either been activated or not. Figure 3.13 illustrates this procedure when the UE detects a lost or weak signal. On such an event, UE suspends all radio bearers and apply default radio configurations. UE moves from the CONNECTED mode to the IDLE mode and starts searching for another cell to camp on. Once the UE finds a suitable cell, UE transmits RRCConnectionRestablishmentRequest to the eNodeB. The newly found suitable cell can be the same old cell or a totally different cell. eNodeB transmits RRCConnectionReestablishment message back to the UE which contains dedicated radio configuration parameters and which the UE uses to configure its sublayers (Table 3.24). The UE reactivates security again, if security has been activated before the beginning of this procedure, and derives the necessary keys for the integrity and ciphering algorithms. Once security is activated, UE moves to the CONNECTED mode and resumes exchange of signalling and data messages using signalling and data radio bearer, respectively.

Finally, the UE transmits the RRCConnectionReestablishment-Complete to the eNodeB to conclude this procedure on SRB1 or SRB1bis.

3.7.12 RRC connection release

If the UE is in the CONNECTED mode, this procedure is used by eNodeB to either release or suspend the connection with the UE. When

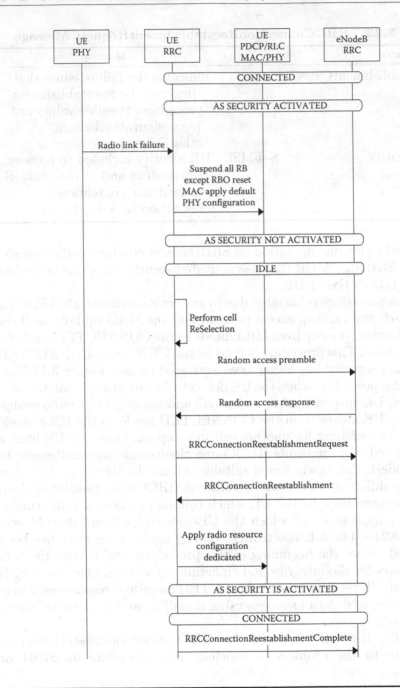

Figure 3.13: Connection re-establishment procedure when AS security is activated.

Table 3.24 RRCConnectionReestablishment Message

Parameter	Size	Meaning
RadioResourceConfig Dedicated	Variable	Includes all dedicated configurations for all sublayers: PDCP, RLC, MAC, and PHY. It also contains also SRB and DRB to be established

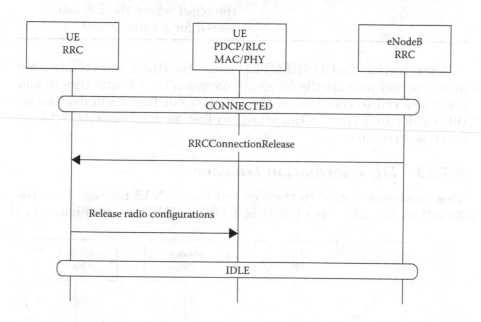

Figure 3.14: Connection release procedure for releasing a connection.

the connection is to be released, UE releases all signalling and radio bearers in addition to all dedicated radio configurations at all sublayers. However, if the RRCConnectionRelease message is for suspending the connection (by setting the releaseCause to rrc-Suspend), UE suspends all signalling and radio bearers, saves AS context, and store resumeIdentity to be used later if the connection is to be resumed by transmitting RRCConnectionResumeRequest as in Table 3.19.

Upon completing this procedure, the UE moves from the CONNECTED mode to the IDLE mode. Figure 3.14 illustrates this procedure. When the UE receives RRCConnectionRelease message to release a connection, it releases all radio bearer including SRB1bis and releases radio configurations in all sublayers.

Table 3.25   RRCConnectionRelease Message

Parameter	Size (Bits)	Meaning
releaseCause	2	Indicates the reason for releasing the RRC connection. Possible values are {rrc-Suspend, other}
resumeIdentity	40	An ID to identify the AS context of the UE
redirectedCarrierInfo	24	Indicates the carrier frequency and the offset where the UE can search for a suitable cell

Upon leaving the CONNECTED mode, the RRCConnectionRelease messsage may indicate the frequency on which the UE first tries to find a suitable cell to camp on. If the UE does not find a suitable cell on this carrier frequency, it can search to find another suitable cell on a different frequency.

3.7.13 DL information transfer

This procedure is used to transfer and tunnel NAS message from the NodeB to the UE if the UE is in CONNECTED mode. Figure 3.15

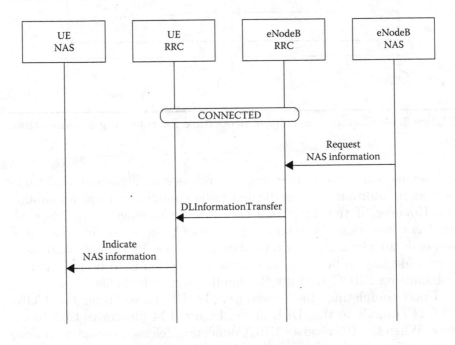

Figure 3.15: DLInformationTransfer procedure.

Table 3.26   DLInformationTransfer Message

Parameter	Size	Meaning
dedicatedInfoNAS	Variable	Carries NAS information piggybacked with this RRC message

illustrates this procedure for tunneling NAS message from the eNodeB to the UE. Once the UE receives the dedicated NAS information, it forwards it to the NAS layer. The content of this message is as shown in Table 3.26.

3.7.14   UL information transfer

This procedure is used to transfer and tunnel NAS message from the UE to eNodeB if the UE is in CONNECTED mode. Figure 3.16 illustrates this procedure for tunneling NAS message from the UE to eNodeB. Table 3.27 shows the content of this message.

If the UE is transmitting RRCConnectionSetupComplete or RRC-ConnectionResumeComplete, the UE can piggyback NAS messages with them and does not need to transmit an ULInformationTransfer message.

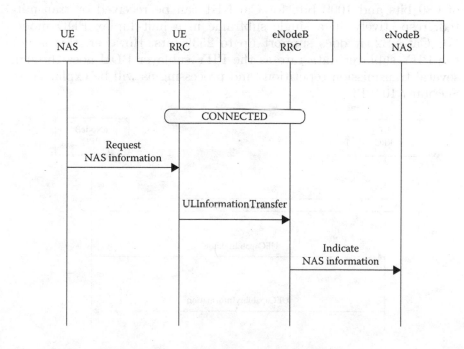

Figure 3.16: ULInformationTransfer procedure.

Table 3.27   ULInformationTransfer Message

Parameter	Size	Meaning
dedicatedInfoNAS	Variable	Carries NAS information piggybacked with this RRC message

3.7.15   UE capability transfer

eNodeB uses this procedure to inquire about the radio access capability of the UE. eNodeB initiates this procedure only if the UE is in CONNECTED mode. Figure 3.17 illustrates this procedure.

eNodeB uses this procedure to query the UE capabilities including AS release version, list of supported bands, support for multiple bearers, support for multi-carrier and multitone operation, maximum number of Robust Header Compression (RoHC) context sessions, and the supported profiles.

This procedure also defines both downlink and uplink capability of UE. The parameters in Table 3.28 are enumerated by the UE. Table 3.28 summarizes all the radio capabilities that are included in the UECapabilityInformation message transmitted to the eNodeB.

Table 3.29 shows the PHY sublayer PDU size that can be received (DL) or transmitted (UL) by a UE in a subframe [19]. A maximum of 680 bits and 1000 bits for Cat-NB1 can be received or transmitted, respectively, in a single subframe in a half-duplex FDD mode. For Cat-NB2, it does support up to 2536 bits. However, this is not the PHY sublayer data rate as the PHY sublayer PDU goes through several transmission repetitions and processing as will be explained in Section 7.10.9.13.

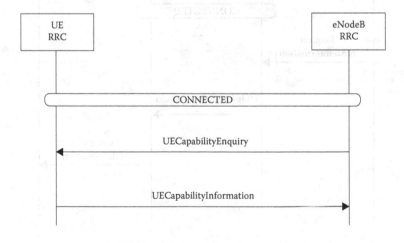

Figure 3.17: Capability enquiry procedure.

Table 3.28 UECapabilityInformation Message

Parameter	Size (Bits)	Meaning
accessStratum Release	4	Indicates the release of the protocol stack. Possible values are {rel13, rel14}
ue-Category-NB	1	If present, defines UE category NB1 as in Table 3.29
multipleDRB	1	If presents, indicates the UE supports multiple DRBs. This parameter is only applicable if the UE supports Data-plane CIoT EPS Optimization. If a UE supports multiple DRBs, the UE shall support two simultaneous DRBs
supportedROHC-Profiles	7	List of supported packet header compression (RoHC) profiles as in Table 4.1
multiTone	1	If presents, indicates the UE supports UL multi-tone transmissions on NPUSCH
multiCarrier	1	If presents, indicates the UE supports multi-carrier operation
multiCarrier-NPRACH	1	If presents, indicates the UE supports NPRACH on non-anchor carrier
twoHARQ-Processes-r14	1	If presents, indicates the UE supports two HARQ processes operation in the DL or UL
supportedBand-List	List	Indicates the list of radio frequency bands supported by the UE
multiCarrier-Paging	1	If presents, indicates the UE supports paging on non-anchor carrier

Table 3.29 Downlink and Uplink Capability

UE Category	Received DL Transport Block Size per TTI	Transmitted UL Transport Block Size per TTI	Duplex
NB1	680	1000	Half-duplex FDD
NB2	2536	2536	Half-duplex FDD

3.7.16 Radio link failure

UE may lose connection with the eNodeB for different reasons such as lost or weak signal, low measured RSRP or low Signal-to-Interference-plus-Noise Ratio (SINR), or unable to decode NPDCCH. A poor or good connection with the eNodeB are both indicated by *out-of-sync* or *in-sync* indications sent from the PHY sublayer to RRC. A good connection (or in-sync) is indicated if the PHY sublayer can decode the NPDCCH successfully a number of consecutive times, N311. A poor connection (or out-of-sync) is defined if the PHY sublayer cannot decode the NPDCCH successfully a number of consecutive times, N310. N311 and N310 can take the values of {1, 2, 3, 4, 5, 6, 8, 10} as configured by eNodeB during the connection establishment procedure.

When the UE detects radio link failure, the UE can decide to initiate the RRC connection re-establishment procedure as explained in Section 3.7.11 or move from the CONNECTED to IDLE mode. When moving to IDLE mode, the UE releases all radio configurations at all sublayers: PDCP, RLC, MAC, and PHY, releases SRBs and DRBs, and performs cell selection to select a new cell to camp on.

3.8   Logical Channels

RRC uses the concept of logical channels to transmit and receive RRC messages to and from eNodeB, respectively. Logical channels used by RRC are control-plane channels. Logical channels are BCCH for receiving broadcast MIB and SIBs, Common Control Channel (CCCH) and Dedicated Control Channel (DCCH) for exchanging RRC messages, and Dedicated Traffic Channel (DTCH) for exchanging data-plane traffic carried on DRBs.

Table 3.30 summarizes the RRC messages and the corresponding radio bearer, logical channel, transport channel, physical channel, and their direction. Logical channels are mapped to transport channels at the MAC sublayer, which are mapped to physical channels at the physical sublayer. The mapping of logical channel to transport and physical channels are explained in Sections 6.1 and 7.8.

3.9   Multi-carrier Support

A UE can support multiple downlink or uplink carrier frequencies to the eNodeB. This is called multi-carrier support. This feature is introduced as load-balancing the large number of NB-IoT devices on different carriers and thus avoiding contention between NB-IoT devices and achieving higher throughput.

A UE in CONNECTED mode is configured, through RRCConnectionReconfiguration message, to an additional non-anchor carrier, for all unicast transmissions. The anchor carrier carries all synchronization

Table 3.30 RRC Messages and Their Channels

Message	Bearer	Logical Channel	Transport Channel	Physical Channel	Direction
MIB	-	BCCH	BCH	NPBCH	DL
SIB1-NB	-	BCCH	DL-SCH	NPDSCH	DL
SIBs	-	BCCH	DL-SCH	NPDSCH	DL
Paging		PCCH	PCH	NPDSCH	DL
RRCConnectionReestablishmentRequest	SRB0	CCCH	UL-SCH	NPUSCH	UL
RRCConnectionRequest	SRB0	CCCH	UL-SCH	NPUSCH	UL
RRCConnectionResumeRequest	SRB0	CCCH	UL-SCH	NPUSCH	UL
RRCConnectionReestablishment	SRB0	CCCH	DL-SCH	NPDSCH	DL
RRCConnectionSetup	SRB0	CCCH	DL-SCH	NPDSCH	DL
DLInformationTransfer	SRB1, SRB1bis	DCCH	DL-SCH	NPDSCH	DL
RRCConnectionReconfiguration	SRB1	DCCH	DL-SCH	NPDSCH	DL
RRCConnectionRelease	SRB1, SRB1bis	DCCH	DL-SCH	NPDSCH	DL
SecurityModeCommand	SRB1	DCCH	DL-SCH	NPDSCH	DL
UECapabilityEnquiry	SRB1, SRB1bis	DCCH	DL-SCH	NPDSCH	DL
RRCConnectionResume	SRB1	DCCH	DL-SCH	NPDSCH	DL
RRCConnectionReconfigurationComplete	SRB1	DCCH	UL-SCH	NPUSCH	UL
RRCConnectionReestablishmentComplete	SRB1, SRB1bis	DCCH	UL-SCH	NPUSCH	UL
RRCConnectionSetupComplete	SRB1bis	DCCH	UL-SCH	NPUSCH	UL
SecurityModeComplete	SRB1	DCCH	UL-SCH	NPUSCH	UL
UECapabilityInformation	SRB1, SRB1bis	DCCH	UL-SCH	NPUSCH	UL
ULInformationTransfer	SRB1, SRB1bis	DCCH	UL-SCH	NPUSCH	UL
RRCConnectionResumeComplete	SRB1	DCCH	UL-SCH	NPUSCH	UL

Table 3.31   Multi-carrier Support

Non-Anchor Carrier	Anchor Carrier		
	Inband	Guardband	Standalone
Inband	Yes	Yes	No
Guardband	Yes	Yes	No
Standalone	No	No	Yes

and system information, while the non-anchor carrier can carry the data transmission, Paging message, random access procedure, or SC-PTM reception. UE uses either the anchor carrier or non-anchor carrier at a time and not both of them simultaneously.

A non-anchor carrier can be configured during RRC connection establishment procedure for unicast transmissions. When a DL non-anchor carrier is configured for the UE, the UE receives data on this carrier frequency. A bitmap can be also provided for this non-anchor carrier that indicates subframes that can be used for receiving data. The non-anchor carrier contains more available subframes for DL since the synchronization and broadcast information are being received on the anchor carrier. UL non-anchor carrier can be also configured for the UE.

If the UE is not configured for a non-anchor carrier, all downlink and uplink transmissions are carried on the anchor carrier only. If the UE is configured for a non-anchor carrier, the UE is either transmitting or receiving on a single carrier only and not simultaneously on all subcarriers. That is, reception and transmission is not done simultaneously and is restricted to only one band for the DL or UL. It is sufficient that the UE has only one transmitter and receiver. Valid anchor and non-anchor carrier combinations are shown in Table 3.31.

3.10   Control-Plane and Data-Plane Cellular IoT (CIoT) Optimization

CIoT EPS optimizations provide a way to support small data or SMS transfer between UE and eNodeB. The UE indicates the support for control-plane CIoT EPS optimization, data-plane CIoT EPS optimization, or S1-U data transfer during RRC Connection establishment procedure, NAS attach request, or NAS TAU request [20, 21]. A UE that supports data-plane CIoT EPS optimization also supports S1-U data transfer.

Control-plane CIoT EPS optimization is used to transport user data (IP, non-IP), SMS messages, or any other data-plane traffic over control-plane via the MME without triggering data radio bearer establishment. The UE NAS uses any of the following NAS procedures to exchange its data-plane messages: Downlink NAS Transport, Uplink NAS Transport, Downlink Generic NAS Transport, and Uplink Generic NAS Transport [20].

Data-plane CIoT EPS optimization is used to change the NAS mode from EMM-IDLE mode to EMM-CONNECTED mode without the need for using the service request procedure [20].

Control-plane CIoT EPS optimization is characterized by the following:

- Enables support of efficient transport of user data (IP, non-IP or SMS) over control-plane without data radio bearer establishment.

- All uplink and downlink NAS messages are piggybacked with RRC messages as shown in Tables 3.26 and 3.27.

- RRC connection reconfiguration is not supported. It is optional for the UE to support RRC Connection Re-establishment procedure.

- Only signalling radio bearer (SRB1bis) is established and no data radio bearer (DRB) is established.

- Only one dedicated logical channel and there is no DTCH supported.

- PDCP sublayer is not used and AS security is not activated.

- UE can support S1-U Transfer.

UE with data-plane CIoT EPS optimization is intended to transfer data-plane traffic without the need for using NAS Service request procedure in order to establish the AS context at the eNodeB. This facilities NB-IoT devices and machine-type communication. Data-plane CIoT EPS optimization is characterized by the following:

- Supports data transfer using data radio bearers. PDCP sublayer is bypassed until it is activated.

- An RRC connection suspend procedure is used when eNodeB releases the RRC connection. eNodeB can request the UE to retain the UE AS context including UE capability while in IDLE mode. This RRC procedure is explained in Section 3.7.12.

■ When moving the UE from IDLE to CONNECTED mode, the RRC connection resume procedure is used. The eNodeB uses and access the UE-stored information to resume the RRC connection. eNodeB uses a Resume ID provided by UE to access the stored UE information. This RRC procedure is explained in Section 3.7.9.

■ When the RRC connection is resumed after it has been suspended; security is continued as activated and Re-keying is not supported in RRC connection resume procedure. The short MAC-I is re-used as the authentication token at RRC connection resume procedure by the UE.

■ A non-anchor carrier can be configured when an RRC connection is established, resumed, reconfigured, or re-established.

■ The NAS protocol [20] can move from EMM-IDLE mode to EMM-CONNECTED mode without the need of Service Request procedure.

SIB2-NB contains information whether the eNodeB supports CIoT EPS optimizations. When the UE receives SIB2-NB at the RRC, it forwards it to the NAS layer at the UE. The UE decides whether to use this optimization or not and transmits this decision during RRC connection establishment procedure.

3.11 Power Saving Mode (PSM)

PSM enables the device to go into a deep sleep mode in order to reduce energy consumption. UE with delay tolerant application or infrequent data transmissions and receptions can use this mode [9, 22].

In PSM, UE decides how long it needs to be in sleep mode. During PSM, the UE is powered-off, remains registered with the network. This allows the UE to avoid re-attach or re-establish PDN connectivity when it becomes active again. During PSM, the UE is not reachable for mobile terminating services and the network is aware of the UE state and avoids paging the UE. If the UE moves to CONNECTED mode, it becomes available for mobile terminating or originating services again.

UE can request to enter the PSM by including the timer, T3324, during an attach or TAU procedure. The EPC grants the PSM to UE by providing a value for timer T3324 in the attach accept or TAU accept procedures.

When the T3324 timer expires, the UE deactivates the AS including PHY, MAC, RLC, PDCP, and RRC sublayers and enters PSM. If the UE was in CONNECTED mode when the T3324 expires, it releases the RRC connection. When the UE enters PSM, it can continue to be in this mode for a time equal to T3412. If UE needs to remain a longer time in PSM than the T3412 time value broadcasted by EPC, it can transmit a second timer, which is an T3412 extended time during the attach or TAU procedures. The maximum duration, including T3412, is about 413 days [21]. Figure 3.18 and Table 3.32 show the format of the two timers T3324 and T3412 [21].

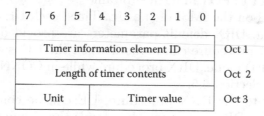

Figure 3.18: T3324 and T3412 extended timer information element.

Table 3.32   T3324 Timer Information Element

Parameter	Meaning
Length of timer contents	Length of the content of the timer information element
Unit	Can be any of the follow for T3324 Timer: 0 0 0 value is incremented in multiples of 2 seconds 0 0 1 value is incremented in multiples of 1 minute 0 1 0 value is incremented in multiples of decihours 1 1 1 value indicates that the timer is deactivated. Can be any of the follow for T3412: 0 0 0 value is incremented in multiples of 10 minutes 0 0 1 value is incremented in multiples of 1 hour 0 1 0 value is incremented in multiples of 10 hours 0 1 1 value is incremented in multiples of 2 seconds 1 0 0 value is incremented in multiples of 30 seconds 1 0 1 value is incremented in multiples of 1 minute 1 1 0 value is incremented in multiples of 320 hours 1 1 1 value indicates that the timer is deactivated
Timer Value	Binary coded timer value

3.12 Discontinuous Reception (DRX) in IDLE Mode

DRX is another procedure by which the UE can conserve energy and battery. In this procedure, UEs need not to monitor the NPDCCH in every subframe to detect whether a Paging message is received. Instead, there is available designated Paging Frames (PF) where each frame can contain one or more Paging Occasion (PO). Both PF and PO are known to the UE and the UE can wake up only on a single PO and detects an NPDCCH that is scrambled with P-RNTI. There is only one PF in each radio frame and one PO in each DRX cycle and the UE monitors only one PO per DRX cycle. DRX cycle has a maximum value of 10.24 seconds.

Within each PF, PO can be in subframe #0, #4, #5, or #9. The PF and PO depend on the IMSI of the UE. Thus, each UE has a different paging occasion. DRX default parameters are provided to the UE in SIB2-NB. DRX procedure can be used when the UE is either in IDLE or CONNECTED mode. DRX procedure while in CONNECTED mode is explained in Section 6.5.

In addition to the DRX procedure, UE can be configured to use extended DRX (eDRX) cycle which extends the sleeping cycle. eDRX cycle length is shown in Table 3.33 [21]. The UE can request the use of DRX/eDRX during attach or TAU procedures by including the DRX/eDRX parameters IEs. EPC accepts the request by transmitting attach or TAU accept messages to the UE. The UE can request to enable both PSM and DRX/eDRX. The EPC decides whether to allow both, or only one, or none of these procedures. However, the UE can request a different combination at each new attach or TAU procedure.

Table 3.33 eDRX Cycle Length

eDRX Cycle
20.48
40.96
81.92 ($\sim$ 1 min)
163.84 ($\sim$ 3 min)
327.68 ($\sim$ 5 min)
655.36 ($\sim$ 11 min)
1310.72 ($\sim$ 22 min)
2621.44 ($\sim$ 44 min)
5242.88 ($\sim$ 87 min)
10485.76 ($\sim$ 175 min)

Chapter 4

Packet Data Convergence Protocol Sublayer

PDCP is a thin sublayer that is used for both control- and data-plane [23]. Its main functionality is to provide integrity and security protections to control- and data-plane PDUs. In particular, PDCP provides the following functionalities:

- Assigning a sequence number to the transmitted PDCP SDU and handling the sequence number of the received PDCP SDU.

- Header compression and decompression, using RoHC protocol, for upper layer packets (e.g., IP layer).

- Ciphering and deciphering of control- and data-plane PDUs.

- Integrity protection and verification for control-plane PDUs only.

- Re-ordering, in-order delivery, and duplicate detection of received SDUs before forwarding to upper layer.

4.1 PDCP Architecture

PDCP architecture is illustrated in Figure 4.1 for control and data RB. Signalling radio bearer, SRB0 and SRB1bis, are used for RRC PDUs

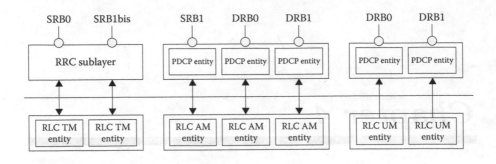

Figure 4.1: PDCP architecture at the UE.

and do not go through the PDCP sublayer. That is, all transmissions and receptions of RRC PDUs on SRB0 and SRB1bis are exchanged between RRC and RLC sublayers without intervention of PDCP. On the other hand, signalling radio bearer, SRB1, goes through PDCP. Any RRC PDU carried over SRB1 can be thus ciphered or integrity protected. Data radio bearer, such as DRB0 or DRB1, also go through PDCP sublayer and can be subject to ciphering.

Each radio bearer that goes through the PDCP has its own PDCP entity that is mapped to either RLC AM or UM. PDCP entity mapped to RLC AM is used for unicast traffic. PDCP entity mapped to RLC UM are used for receiving multicast traffic on SC-MCCH or SC-MTCH only and not used for unicast traffic. Each PDCP entity means that each one has its own state, state variables, and operation independent of the other entities.

UE that only supports control-plane CIoT EPS optimization, as defined in [20], has its PDCP sublayer bypassed. For an NB-IoT UE that supports both control-plane CIoT EPS optimization and data-plane CIoT EPS optimization, as defined in [20], PDCP is also bypassed (i.e., not used) until AS security is activated.

4.2 RRC Configuration Parameters

RRC signals configuration parameters to PDCP sublayer to configure the integrity and ciphering algorithms, packet header compression algorithm, or DRBs as shown in Table 4.1. The PDCP configuration parameters are received by RRC from eNodeB during RRC connection establishment procedure as explained in Section 3.7.7. Data-plane packets received from upper layer (e.g., TCP/IP) and received by PDCP sublayer are mapped to one of the DRB IDs indicated by drb-Identity.

Table 4.1   RRC Configuration Parameters for PDCP Sublayer

Parameter	Size (Bits)	Meaning
esp-BearerIdentity	4	Indicates the EPS bearer ID as explained in Chapter 8
drb-Identity	5	Indicates the DRB ID used for each DRB established
cipheringAlgorithm	4	Cipher algorithm to be used for ciphering signalling and data RB as in Table 3.16
integrity-ProtAlgorithm	4	Integrity algorithm to be used for protecting signalling RB as in Table 3.17
discardTimer	3	Indicates the discard timer in milliseconds. Possible values are {ms5120, ms10240, ms20480, ms40960, ms81920, infinity}
headerCompression	10	If presents, indicates the packet header compression (RoHC) profile used with a PDCP entity. Possible values as in Table 4.3

Integrity and security algorithms are signalled to PDCP during the RRC security activation procedure as explained in Section 3.7.8.

SRB1 uses the integrity and security parameters but not the other parameters. Since SRB0 and SRB1bis do not go through PDCP sublayer, they do not possess PDCP configurations.

4.3   PDCP Entity

Figures 4.2 and 4.3 illustrate the structure for a PDCP entity that is used for either the control-plane or data-plane, respectively. In Figure 4.2, a control-plane PDU or a signalling SRB PDU is first assigned a Sequence Number (SN), integrity protected and ciphered, and the PDCP header is added before it is transmitted to RLC sublayer.

In Figure 4.3, a data-plane PDU or data DRB PDU is not integrity protected, instead it is ciphered after an SN is assigned to it; PDCP header is added and finally transmitted to lower sublayers.

A PDCP PDU consists of a header and an SDU. Both are multiple of bytes. The PDCP PDU consists of the PDCP SDU and PDCP header as shown in Figure 4.4. The PDCP PDU has a maximum size of 1600 octets.

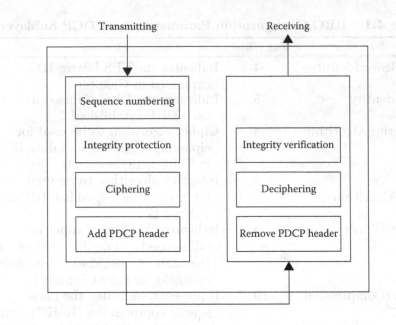

Figure 4.2: PDCP entity for control-plane (Signalling RB).

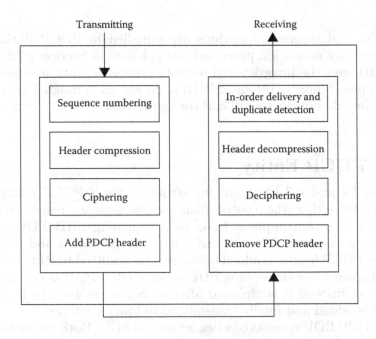

Figure 4.3: PDCP entity for data-plane (Data RB).

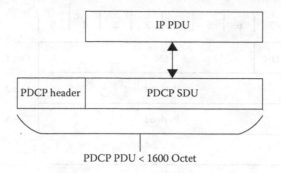

Figure 4.4: PDCP SDU and PDU.

PDCP PDU can be of data type or control one. The first means that the PDCP SDU carries either a signalling or data PDU (i.e., control-plane or data-plane PDU), while the latter means that it carries control information such as Interspersed RoHC feedback packet.

Figure 4.5 shows the PDCP data PDU format that carries a signalling PDU. Data PDU uses a sequence number length of 5 bits and carries a signalling PDU mapped to an SRB. The MAC-I field is the message authentication code calculated by the integrity algorithm and appended to the PDCP SDU.

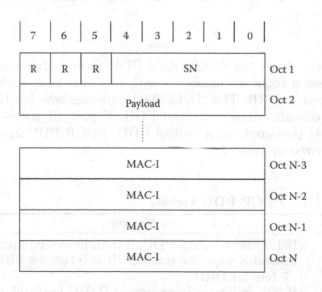

Figure 4.5: PDCP PDU format for Signalling PDU (5 bits SN).

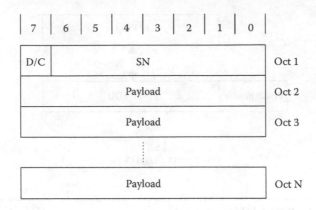

Figure 4.6: PDCP PDU format for data PDU (7 bits SN).

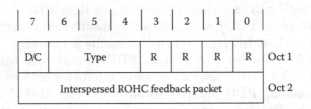

Figure 4.7: PDCP PDU format for interspersed RoHC feedback control SDU.

Figure 4.6 shows the PDCP data PDU format that carries data PDU. It uses a sequence number length of 7 bits and carries a data PDU mapped to DRB. The "D/C" field indicates whether the PDCP data SDU contains data or control PDU. Figure 4.7 shows a PDCP control PDU that contains a control PDU. PDCP PDU header fields are summarized in Table 4.2.

Table 4.2 PDCP PDU Fields

Field	Meaning
D/C	If 1, indicates data PDU; if 0, indicates control PDU
SN	Indicates sequence number. It is 5 bits for SRB and 7 bits for DRB
Type	If 001, indicates Interspersed RoHC feedback packet
R	Reserved

4.4 Ciphering and Deciphering

Ciphering and deciphering refers to the process of encrypting or decrypting the PDCP PDU. Ciphering is activated by RRC sublayer when receiving RRC SecurityModeCommand PDU. The parameters used for ciphering and deciphering include the following [24]:

- **KEY**: Both the keys, K_{RRCenc} and K_{UPenc}, are driven by RRC and used for ciphering signalling or data-plane PDU, respectively. Keys are 128 bit long.

- **BEARER**: 5-bit bearer ID.

- **COUNT**: 32-value that is the concatenation of HFN and PDCP PDU SN.

- **DIRECTION**: 0 for uplink and 1 for downlink.

Ciphering is applied only to PDCP SDU which includes a data PDU (a control-plane or data-plane PDU) and does not apply to PDCP control SDU. Ciphering is also applied to the MAC-I field which carries the message authentication code of the integrity algorithm.

Figure 4.8 shows the EPS Encryption Algorithm (EEA) used with the input parameters: KEY, BEARER, COUNT, and DIRECTION. The Length input parameter is the length of the required keystream block which is typically set equal to the length of the PDCP SDU. Once

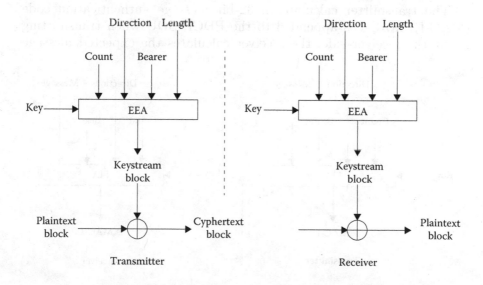

Figure 4.8: Ciphering and deciphering block diagram.

PDCP SDU is ciphered, it is transmitted over the air and deciphered on the other side using the same input parameters.

4.5   Integrity Protection and Verification

Integrity refers to the process of adding a hash value to the PDCP PDU to verify its integrity and detect any tampering with the PDU. Integrity is activated by RRC sublayer because of SecurityModeCommand procedure. The parameters used for integrity protection and verification includes the following [24]:

- **KEY**: The key, K_{RRCint}, that is driven by RRC and used for integrity protecting signalling PDU. Key is 128 bits long.

- **BEARER**: 5-bit bearer ID.

- **COUNT**: 32-value that is the concatenation of HFN and PDCP PDU SN.

- **DIRECTION**: 0 for uplink and 1 for downlink.

Integrity protection is applied only to PDCP header and payload (PDCP SDU) of a control-plane PDU before ciphering. Figure 4.9 illustrates the EPS Integrity Algorithm (EIA) that authenticates the integrity of a signalling PDU.

The transmitter calculates a 32-bit message authentication code (MAC-I). MAC-I is appended to the PDCP SDU when transmitting it. At the receiver side, the receiver calculates the expected message

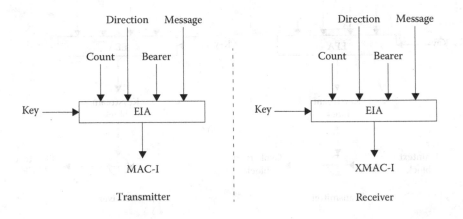

Figure 4.9: Integrity block diagram.

authentication code (XMAC-I) for the received PDCP PDU the same way it was calculated at the sender. If MAC-I and XMAC-I matches, the PDCP PDU passes the integrity verification; otherwise, the PDCP PDU fails the integrity verification.

4.6   Header Compression and Decompression

Each PDCP entity supports header compression and decompression according to the RoHC framework. There are different compression algorithms or profiles depending on which protocols the network layer is using. Table 4.3 summarizes all header compression and decompression profiles supported by UE. Header compression applies only to data radio bearer but not signalling radio bearers.

RRC sublayer configures the PDCP and each PDCP entity to which profile ID to be used for its header compression algorithm.

Header compression or decompression applies only to data-plane PDUs. That is, before ciphering, network layer packets (e.g., TCP/IP) mapped to a single PDCP PDU gets compressed.

Compression algorithm also can generate another SDU called interspersed RoHC feedback packet required for the operation of the compression algorithm. The feedback packet is encapsulated in PDCP PDU without an SN as in Figure 4.7. This PDCP PDU is not ciphered. On the receiver side, this PDCP PDU is neither deciphered nor decompressed. The PDCP header fields are as in Table 4.2. UE can use the header compression algorithms defined in Table 4.3 for both uplink and downlink.

Table 4.3   Header Compression and Decompression Algorithms

Profile ID	Transport/Network Layer Protocols	Reference
0x0000	No compression	RFC 5795
0x0002	UDP/IP	RFC 3095, RFC 4815
0x0003	ESP/IP	RFC 3095, RFC 4815
0x0004	IP	RFC 3843, RFC 4815
0x0006	TCP/IP	RFC 6846
0x0102	UDP/IP	RFC 5225
0x0103	ESP/IP	RFC 5225
0x0104	IP	RFC 5225

4.7 PDCP Transmission

4.7.1 Data and signalling radio bearer transmission on uplink

PDCP sublayer receives packets from upper layer (i.e., PDCP SDU), assigns an SN to it, integrity- and cipher-protecting those packets and forward them to RLC sublayer.

For the PDCP entity that transmits a PDCP PDU to RLC, it maintains a discard timer for each PDCP PDU. In addition, there is a state variable, Next_PDCP_TX_SN, that is initially set to zero. Upon transmitting the data PDU to lower layer, the discard timer associated with the PDU is started and PDCP sets the SN of outgoing PDCP PDU to Next_PDCP_TX_SN, perform header compression (for DRB only), integrity protection (for SRB only), ciphering, and increment Next_PDCP_TX_SN. This is illustrated in Figure 4.10. The SN is 7 bits (for DRB) and 5 bits (for SRB) and if Next_PDCP_TX_SN exceeds 127 (for DRB) or 32 (for SRB), the counter is set to zero and TX_HFN is incremented by one. The state variables used for a PDCP entity is summarized in Table 4.4.

When a PDCP PDU is transmitted to RLC, a timer, discardTimer, is started so that if no ACK or NACK is received from RLC sublayer (as a result of receiving it from the eNodeB), the PDCP PDU is discarded and not retransmitted again to RLC. This timer is stopped if the PDCP entity received an ACK for the transmitted PDCP PDU.

A COUNT is a 32-bit state variable that consists of the concatenation of the SN and Hyper Frame Number (HFN). COUNT is used as a parameter during integrity and ciphering algorithms. The number of bits for the SN is different for signalling and data radio bearer (Figure 4.11). SN is 5 bits for signalling while it is 7 bits for data. The size of the HFN in bits is either 27 bits (if SN is 5 bits) or 25 bits (if SN is 7 bits) (Figure 4.12).

Table 4.4   State Variables Maintained by a PDCP Entity when Transmitting Data on Uplink

State Variable	Initial Value	Meaning
Next_PDCP_TX_SN	0	Indicates SN to be assigned to the next PDCP PDU to be transmitted for a given PDCP entity
TX_HFN	0	Indicates the HFN used for generation of the COUNT value for a given PDCP entity

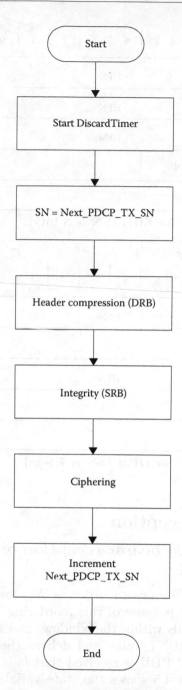

Figure 4.10: PDCP PDU transmission on uplink.

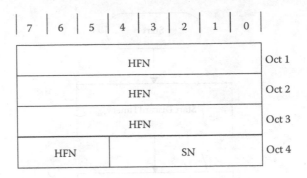

Figure 4.11: COUNT for SRB (SN is 5 bits).

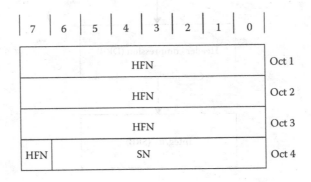

Figure 4.12: COUNT for DRB (SN is 7 bits).

4.8 PDCP Reception

4.8.1 *Data radio bearer reception on downlink RLC AM*

Each PDCP entity maintains a reordering Window which is always half of the SN space. The purpose of this reordering Window is to receive PDCP PDUs that falls within the window, and then to reorder them according to the COUNT value and deliver them in-order to upper layer. When a PDCP PDU is received that falls outside the window, it is discarded. Table 4.5 shows the state variables used by a PDCP entity.

Figure 4.13 explains the concept of reordering Window and state variables. PDCP SN wraps around and if a received PDCP SN falls within the window, it will be further processed; otherwise, it

Table 4.5 **State Variables of a PDCP Entity for DRB Mapped to RLC AM**

State Variable	Initial Value	Meaning
Last_Submitted _PDCP_RX_SN	127	The SN of PDCP PDU last submitted to upper layer (IP layer) for a given PDCP entity
Next_PDCP_RX_SN	0	The SN of next expected PDCP PDU to be received for a given PDCP entity
PDCP SN	0	The SN of PDCP PDU that is received for a given PDCP entity
RX_HFN	0	HFN used for the generation of the COUNT value for the received PDCP PDU for a given PDCP entity

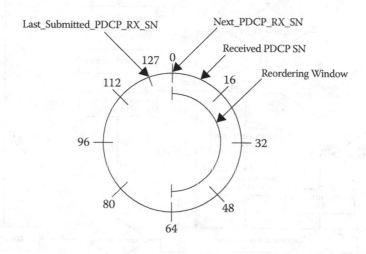

Figure 4.13: Reordering window and PDCP entity (7 bits SN) state variables.

is discarded. If the PDCP is not discarded, the RX_HFN and state variables are updated according to Figure 4.14.

Figure 4.14 shows the flow chart of receiving a PDCP PDU on the downlink for an SN space consists of 7 bits (an SN from zero to 127) and reordering Window half of the SN (63). If the SN of the received PDCP PDU is outside the reordering Window, the PDCP PDU is discarded. Before the PDCP PDU is discarded, it is still deciphered and

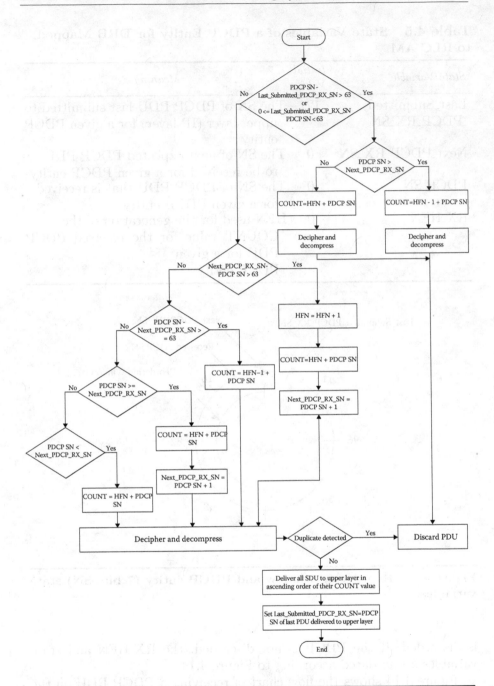

Figure 4.14: Data reception on downlink RLC AM.

decompressed in order to keep the state of the cipher and decompression algorithm. RX_HFN is determined and associated with the received PDCP PDU to determine the COUNT value used for deciphering and integrity verification.

4.8.2 Data radio bearer reception on downlink RLC UM

DRB used with RLC UM are used for receiving multicast data and signalling PDUs on SC-MTCH or SC-MCCH, respectively. This is because RLC UM is only supported for SC-MCCH and SC-MTCH. PDCP entity mapped to RLC UM does not use reordering Window. If such a traffic is received, the SN of the received PDCP is compared to a state variable Next_PDCP_RX_SN. The traffic is deciphered and the header is decompressed (if the DRB is configured by RRC for header compression) and finally forwarded to the upper layer. Figure 4.15 shows the flow chart for receiving PDCP PDUs on downlink RLC UM.

4.8.3 Signalling radio bearer reception on downlink

Signalling PDUs reception on the downlink is shown in Figure 4.16 and does not have a reordering Window. If a PDCP PDU is received, it is deciphered and its integrity is verified. The HFN is chosen depending on whether its SN is greater or less than the expected SN of PDCP PDU to be received next. COUNT is calculated from the HFN and the received SN. Integrity and ciphering algorithms can be run to decipher and verify the integrity of the received PDCP PDU.

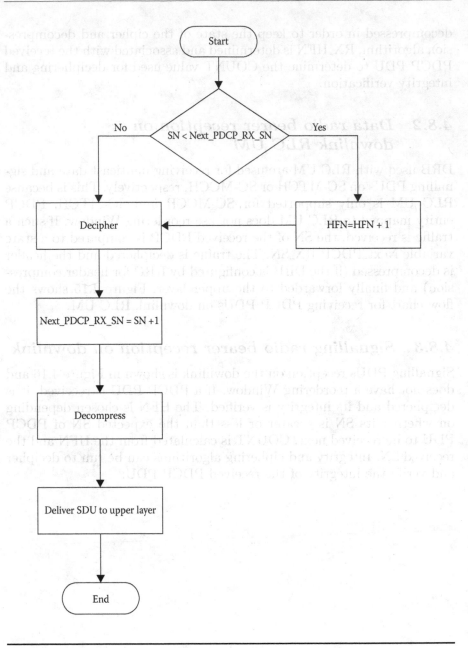

Figure 4.15: Data reception on downlink RLC UM.

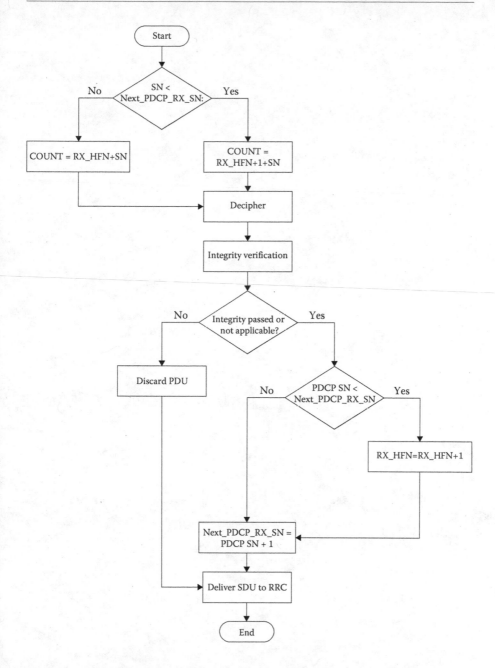

Figure 4.16: Signalling reception on downlink.

Chapter 5

Radio Link Control Sublayer

RLC is an important sublayer in the LTE™ NB-IoT protocol stack. It is responsible for the reliable and guaranteed transfer of control- and data-plane PDUs to the receiver side. The RLC sublayer provides the following functionalities [25]:

- Reliable transfer of RLC PDUs with the other peer.

- Flow control and error handling through ARQ.

- Segmentation or concatenation of RLC SDUs.

- Reassembly, reordering, or duplicate detection of received RLC PDUs.

5.1   RLC Architecture

Similar to the PDCP sublayer architecture, each signalling or data radio bearer has and is associated with a single RLC entity. A transmitter and a receiver both have a peer-to-peer RLC entity at each of them that is exchanging RLC PDUs. The payload of an RLC PDU is typically the PDCP or RRC PDU passed from or to PDCP or RRC sublayer, respectively. Each RLC entity has its own state variables and independent of the operation of others.

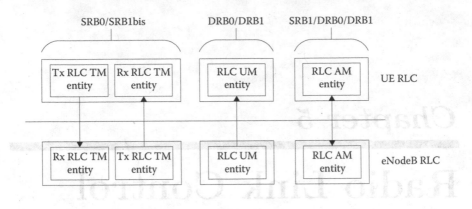

Figure 5.1: RLC architecture at the UE.

RLC entity can be one of the three modes: Transparent Mode (TM), Unacknowledgment Mode (UM), or Acknowledgement Mode (AM). RLC entity of TM is an unidirectional entity which means that the entity is for one direction only, either transmitting (i.e., UL) or receiving (i.e., DL). RLC mode, Unacknowledged Mode (UM), is used only to receive multicast traffic. RLC UM is used for SC-MCCH and SC-MTCH traffic. RLC entity of an AM mode is bidirectional which means that the RLC entity is used for both transmitting and receiving. RRC sublayer configures each RLC entity as a transmitting TM, receiving TM, receiving UM, or AM. RRC also associates each RLC entity with its signalling or radio bearer.

Figure 5.1 illustrates the RLC architecture for TM, UM, and AM. TM are used for SRB0 and SRB1bis while AM is used for SRB1 and DRBs. UM is used only for DRBs.

5.2 RRC Configuration Parameters

RRC sends configuration parameters to RLC to configure the RLC entities used for SRBs or DRBs as shown in Table 5.1. The RLC configuration parameters are received by RRC from eNodeB during RRC connection establishment procedure as explained in Section 3.7.7.

5.3 RLC Entity

5.3.1 Transparent mode

Figure 5.2 illustrates the architecture of an RLC entity with TM. An RRC PDU received from RRC sublayer is queued for transmission in

Table 5.1   RRC Configuration Parameters for RLC Sublayer

Parameter	Value	Meaning
logical-ChannelIdentity	[3 10]	Indicates the logical channel identity used for both UL and DL for a DRB
t-PollRetransmit	ms250, ms500, ms1000, ms2000, ms3000, ms4000, ms6000, ms10000, ms15000, ms25000, ms40000, ms60000, ms90000, ms120000, ms180000	Indicates a periodic interval of time, in milliseconds, where the RLC transmitter polls the receiver to send an RLC STATUS PDU
maxRetxThreshold	t1, t2, t3, t4, t6, t8, t16, t32	Indicates the maximum number of RLC SDU retransmissions
enableStatus-ReportSN-Gap	True	If enabled, indicates that RLC is to transmit an RLC STATUS PDU when it detects a failure in receiving an RLC PDU
t-Reordering	ms0, ms5, ms10, ms15, ms20, ms25, ms30, ms35, ms40, ms45, ms50, ms55, ms60, ms65, ms70, ms75, ms80, ms85, ms90, ms95, ms100, ms110, ms120, ms130, ms140, ms150, ms160, ms170, ms180, ms190, ms200, ms1600	Indicates the time, in milliseconds, to trigger reordering of the buffered RLC PDUs and delivering them to PDCP

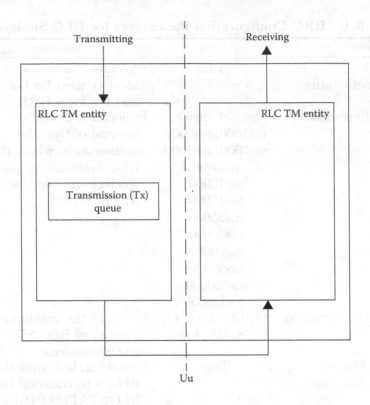

Figure 5.2: RLC entity of TM.

the transmission buffer until an uplink opportunity for transmission is indicated by the MAC sublayer. The transmission opportunity must have a size that can accommodate the RLC SDU queued in the buffer. TM does not append an RLC header to the SDU and RLC PDU is transmitted to MAC sublayer the same as it was received from the RRC sublayer. RLC SDU can be of a variable size and is never segmented nor concatenated. When an RLC PDU is transmitted, it is removed from the transmission buffer.

On the receiving side, if an RLC PDU is received from the MAC sublayer, it is forwarded to the RRC sublayer as it is without any RLC processing. RLC SDU in this mode are not acknowledged by the receiving peer.

SRB0 and SRB1bis are mapped to RLC TM entity. RLC TM entity exchanges downlink and uplink traffic on DL or UL CCCH (SRB0), DL or UL DCCH (SRB1bis), DL PCCH, or DL BCCH.

5.3.2 Unacknowledgement mode

Figure 5.3 illustrates the architecture of an RLC entity for UM. RLC UM is used only by a UE for receiving multicast traffic on SC-MCCH and SC-MTCH. The multicast traffic is always flowing from eNodeB to UE, and thus, the UE acts always as a receiver and not as an RLC UM transmitter.

RLC PDU received by UE in this mode can contain one of the three SDUs: a single RLC SDU, concatenated RLC SDUs, or a segment of an RLC SDU.

When an RLC PDU is received, the receiver detects if it is a duplicate, reorders them, reassembles the received RLC PDUs and delivers them to the PDCP sublayer. The RLC receiver does not transmit ACK/NACK nor does it require re-transmission of missing RLC PDUs

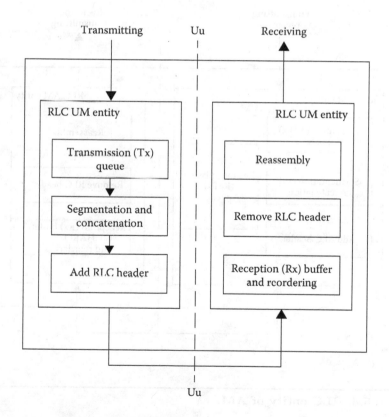

Figure 5.3: RLC entity of UM.

from the transmitter[1]. An RLC UM entity receives downlink traffic on
SC-MCCH and SC-MTCH.

5.3.3 Acknowledgement mode

Figure 5.4 illustrates the architecture of an RLC entity for AM. An
RLC PDU in this mode can be one of the four SDUs: a single RLC
SDU, a concatenated RLC SDUs, a segment of an RLC SDU, or a
segment of an RLC SDU segment.

If an RLC SDU is received from the PDCP sublayer, it is queued
in the transmission buffer until a transmission opportunity is received
from the MAC sublayer. The transmission opportunity can be of any
size. That is, if the transmission opportunity is smaller than the queued
RLC SDU, the RLC SDU is segmented and the RLC PDU segment
is transmitted to the MAC sublayer. If the transmission opportunity

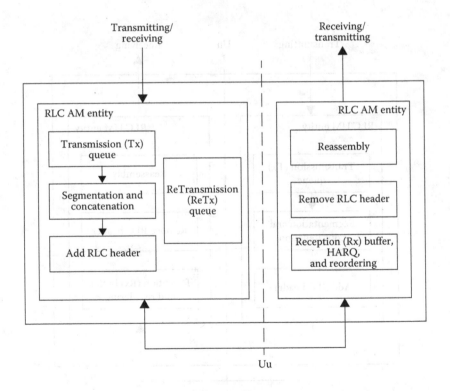

Figure 5.4: RLC entity of AM.

---

[1]In a multicast traffic, transmitted packets are not re-transmitted and neither ACKed
by the receiver.

is big enough to hold one or more RLC SDU, a single RLC SDU or a number of RLC SDUs concatenated together are transmitted in a single RLC PDU.

The transmission opportunity must be of a size big enough to hold the RLC header in addition to the RLC SDU(s) or an RLC SDU segment. Whether an RLC SDU or an RLC SDU segment is transmitted to the MAC sublayer, an RLC header is added to the RLC SDU.

The transmitter does support ARQ. An RLC PDU that is transmitted to the MAC sublayer is not removed and instead is moved from the transmission queue to a retransmission queue. If this RLC PDU is to be retransmitted, it is pulled from the retransmission queue and either retransmitted as it is or segmented according to the transmission opportunity size indicated by the MAC sublayer. An RLC SDU or an RLC SDU segment can be further segmented or multiple RLC SDUs can be further concatenated depending on the size of the uplink transmission opportunity.

An RLC AM entity exchanges downlink and uplink traffic on DL or UL DCCH (SRB1), DL or UL DTCH (DRBs).

On the receiver side, when an RLC PDU is received, the receiver detects if it is a duplicate, reorders them, reassembles the received RLC PDUs and delivers them to the PDCP sublayer. If the receiver detects missing RLC PDUs, it can transmit a special control RLC PDU called RLC STATUS PDU which NACKs those missed RLC PDUs and polls the transmitter to retransmit them.

5.4   RLC PDU Format

RLC PDU is a variable size PDU that consists of multiple octets. The transmission queue or retransmission queue contains RLC SDU(s) and an RLC header is added to the SDU to form the RLC PDU if an uplink transmit opportunity is available for transmitting the RLC SDU. Detailed description of an RLC PDU for TM, UM, and AM will follow in the next sections.

5.5   RLC Transmission and Reception

5.5.1   RLC TM

PDCP PDU coming from the RRC sublayer are queued into the transmission (Tx) queue in the RLC entity of TM in a form of an RLC SDU. An RLC SDU is typically a signalling RRC PDU. When the RLC entity

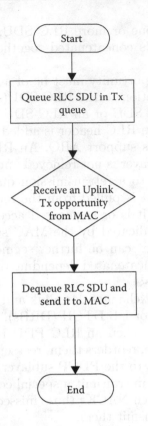

Figure 5.5: RLC TM transmitter side.

receives a MAC indication for an uplink transmission opportunity that can fully accommodate the queued RLC SDU, RLC entity can transmit a single RLC SDU in a single RLC PDU. There is no RLC PDU header added and the RLC PDU is passed to the MAC sublayer for transmission. Figure 5.5 summarizes the TM transmit side processing. Figure 5.6 shows the RLC PDU for TM where the RLC PDU does not have an RLC header and only a payload.

5.5.2 *RLC UM*

An RLC UM entity at the UE acts as a receiver only and not a transmitter as it receives multicast traffic only. RLC UM PDU consist of 5 bits SN which provides an SN space from 0 to 31 expressed as [0 31]. When the transmitter exhausts the SN space (i.e., it reaches SN #31), it wraps the SN around to 0 again.

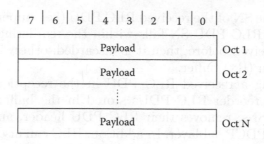

7	6	5	4	3	2	1	0	
Payload								Oct 1
Payload								Oct 2
Payload								Oct N

Figure 5.6: RLC PDU format for TM.

For the RLC receiver, in order to process RLC PDUs that can arrive out-of-order or are duplicates to previously received RLC PDUs, RLC employs a sliding window protocol. This sliding window is implemented as a reorderding Window scheme in RLC UM. Figure 5.7 illustrates the concept of a reordering Window. The reordering Window has an upper edge and lower edge representing a range of SNs. The reordering Window slides to a new window when a new RLC PDU is received with SN that falls outside the reordering Window. The reordering Window acts as a window where all received RLC PDUs that fall within the window are stored and later are reordered and delivered in-order to PDCP sublayer.

When a new RLC PDU, with SN #x, is received by the RLC, it is checked if its SN falls within the reordering Window or falls outside the reordering Window. If it falls outside the reordering Window, the RLC PDU is stored in the reception (Rx) buffer and the reordering Window slides to a new position where the upper edge points to the

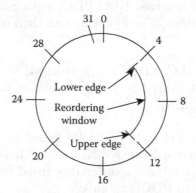

Figure 5.7: Reordering window as a sliding window.

SN following the SN of the received RLC PDU (i.e., upper edge is $x+1$). If the received RLC PDU SN falls within the reordering Window and it has been received before, then it is discarded; otherwise, it is stored in the reception (Rx) buffer.

After storing a received RLC PDU in the reception (Rx) buffer, RLC starts to reorder RLC PDUs stored in the buffer in ascending order of their SNs, removes their RLC PDU header, and delivers the RLC SDUs to PDCP sublayer. In addition, RLC can also start a timer, t-Reordering timer, which upon its expiry, triggers reordering of RLC PDUs and delivering them to PDCP sublayer.

Since RLC UM is used for receiving multicast traffic and in order to keep its implementation in NB-IoT devices simple, 3GPP™ has chosen to set both the values of the reordering Window and t-Reordering timer to zero [15]. The SN field of the RLC PDU is 5 bits. As the value of the reordering Window is zero, this means that both the upper and lower edge have the same SN value and hence any received RLC PDU is considered as falling outside the reordering Window and is to be stored in the reception (Rx) buffer. In addition, since the t-Reordering timer has an interval of zero, this means that upon storing an RLC PDU in the reception (Rx) buffer, RLC can start reordering the buffered RLC PDUs in ascending order of their SNs, strips their RLC PDU headers, and delivers them to PDCP sublayer[2].

RLC UM possesses several state variables. Table 5.2 shows the state variables used by an RLC UM entity at the receiver. VR(UH) represents both the upper and lower edge of the reordering Window as the reordering Window has a size equals to zero. The flow chart for updating the state variables upon reception of an RLC PDU is shown in Figure 5.8.

Figure 5.8 explains the RLC UM receiver behavior. Since reordering Window is zero, any received RLC PDU, with SN = x, is stored into the reception (Rx) buffer. The following can then occur in the following order:

■ All buffered RLC PDUs are reordered, striped from the RLC header, and delivered to PDCP sublayer in ascending order of the RLC PDU SN.

■ If x equals to VR(UR), VR(UR) is set to the SN of the RLC PDU that has not been received yet. All buffered RLC PDUs, with SN<VR(UR), are reordered, striped from the RLC header, and delivered to PDCP sublayer.

---

[2]The RLC UM behavior complies with mutlicast traffic characteristics where packets are transmitted only once and not ACKed.

Table 5.2   RLC UM State Variables

State Variable	Usage	Initial Value	Meaning
UM_Window_Size	Window Size	0	It is zero for SC-MCCH, SC-MTCH
VR(UH)	Highest received	0	Indicates the value of the SN following the SN of the RLC PDU with the highest SN among received RLC PDUs, and it serves as the higher edge of the reordering window
VR(UR)	Receive	0	Indicates the value of the SN of the earliest RLC PDU that is still considered for reordering
VR(UX)	t-Reordering	0	Indicates the value of the SN following the SN of the RLC PDU which triggered t-Reordering

■ If VR(UH) is greater than VR(UR), VR(UX) is set equal to VR(UH), VR(UR) is set to the SN≥VR(UX), and all buffered RLC PDUs, with SN<VR(UR), are reordered, striped from RLC header, and delivered to PDCP sublayer.

Figures 5.9 and 5.10 show the RLC PDU format for UM. The RLC PDU fields are as shown in Table 5.5.

5.5.3   Example of an RLC UM transmitter and receiver

To clarify the UM transmitter and receiver operations by an example, consider the example in Figure 5.11. In this example, the RLC SN field is 5 bits in length and reordering Window size is zero.

The transmitter has sent a number of RLC PDUs where their SNs are incremented. At the receiver, it initially has its state variables VR(UH), VR(UR), and VR(UX) are all set to zero. Receiver receives only RLC PDU with SN #0, #1, #2, #5 while those RLC PDU with SN #3 and #4 are lost over the air and have not been received at the receiver.

Since RLC PDUs with SN equal to 0,1,2,5 all fall outside the reordering Window, they are all accepted and stored in the reception (Rx) buffer. Each received RLC PDU follows the algorithm in Figure 5.8. The values of state variables after receiving each RLC PDU are shown.

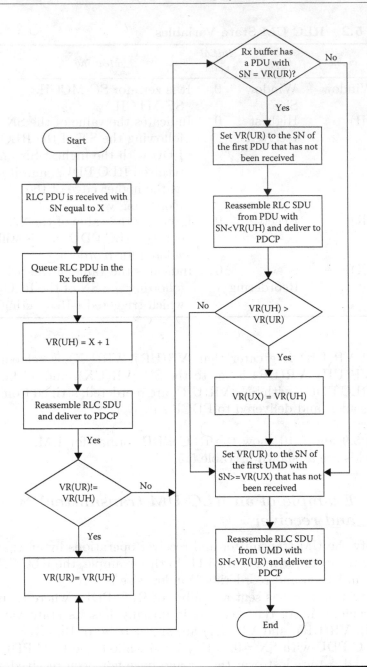

Figure 5.8: RLC UM receiver side.

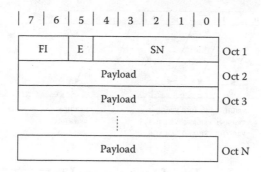

Figure 5.9: RLC PDU for a single RLC SDU or an RLC SDU segment (5 bits SN).

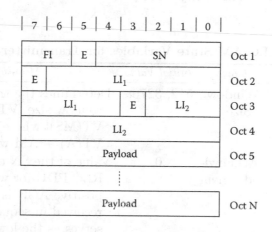

Figure 5.10: RLC PDU for UM when concatenating multiple RLC SDUs (5 bits SN).

After finishing receiving RLC PDU with SN #5, state variables are updated to VR(UH) = VR(UR) = 6, VR(UX) = 0. After receiving an RLC PDU, it is reordered and reassembled, stripped from its RLC header, and delivered to PDCP sublayer.

5.5.4 RLC AM transmitting side

RLC AM entity maintains a number of state variables for both transmitting and receiving operations as in Table 5.3. RLC AM entity uses a sliding window protocol in order to manage flow control, segmentation

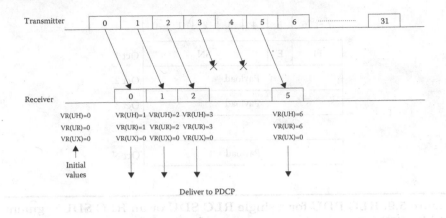

Figure 5.11: RLC UM entity example.

Table 5.3 RLC AM State Variables for Transmitter

State Variable	Usage	Initial Value	Meaning
AM_Window_Size	Window Size	512, 32768	Determines the transmit window size [VT(A) VT(MS)[where VT(MS)= VT(A) + AM window size
VT(A)	Acknowledgement	0	Value of the SN of the next RLC PDU for which a positive ACK is to be received in-sequence. It serves as the lower edge of the transmitting window
VT(MS)	Maximum send		Set to VT(A) + AM_Window_Size, and it serves as the higher edge of the transmitting window
VT(S)	Send	0	Value of the SN to be assigned for the next newly generated RLC PDU
POLL_SN	Poll send state	0	Holds the value of VT(S)−1 upon the most recent transmission of an RLC data PDU with the poll bit set to 1

and reassembly, and in-order delivery. Both transmitter and receiver maintain a window of size AM_Window_Size.

The transmitter maintains a window size, AM_Window_Size, that has a lower edge and an upper edge equal to VT(A) and VT(MS), respectively (i.e., [VT(A) VT(MS)[). When the RLC receives an indication from the MAC sublayer about an uplink opportunity to transmit uplink RLC PDU, RLC schedules RLC SDU(s) for transmission in the following strict order:

- RLC Control PDU.

- Retransmission of an RLC SDU from ReTransmission (ReTx) queue.

- Transmission of an RLC SDU from Transmission (Tx) queue.

Transmitting a new RLC SDU from the Tx queue is illustrated in the flow chart in Figure 5.12. PDCP PDU coming from the PDCP sublayer are queued into the Tx queue in the RLC entity. When the RLC entity receives a MAC indication for an uplink transmission opportunity and depending on the size of this uplink transmission opportunity, RLC entity can transmit a single RLC SDU, concatenation of multiple RLC SDUs, or a segment of an RLC SDU. In all cases, RLC assigns an SN to the RLC PDU, containing the RLC SDUs, equal to VT(S) and increments VT(S). VT(S) must be in the range of [VT(A) VT(MS)[; otherwise, this RLC PDU is not scheduled for transmission. If VT(S) becomes equal to VT(MS), the transmit window becomes stalled and no further transmission of an RLC PDU is possible. When an RLC PDU is to be transmitted, a header is added and the RLC PDU is passed to the MAC sublayer for transmission.

The transmitter can receive the ARQ feedback from the receiver. That is, transmitter receives an ACK or NACK for its transmission in the form of a control RLC PDU called RLC STATUS PDU. The flow chart illustrating the receiver behavior when an RLC STATUS PDU is received is shown in Figure 5.13. If an RLC PDU with its SN equal to VT(A) is positively ACKed by the receiver, the transmitter window, [VT(A) VT(MS)[, slides to a new value where VT(A) is set to the smallest SN of an RLC PDU that is yet to be ACKed.

Figure 5.14 illustrates the operation of the transmitter sliding window, [VT(A) VT(MS)[, and the state variable, VT(S). Upon transmitting an RLC PDU, the SN of the RLC PDU is set equal to VT(S) and VT(S) is incremented by one. The sliding window only slides when an ACK is received for an RLC PDU with SN equal to the lower edge of the sliding window, VT(A).

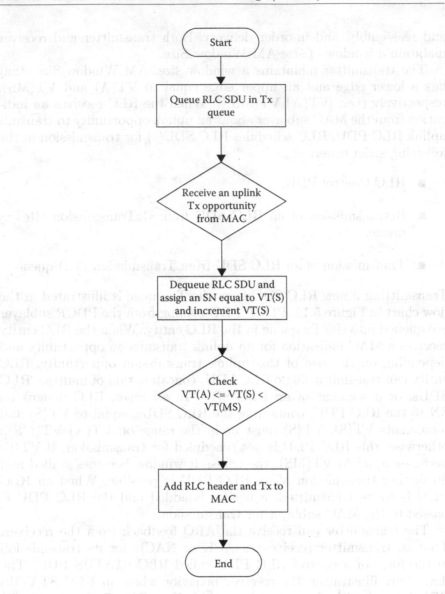

Figure 5.12: RLC AM transmitter side.

5.5.5 RLC AM retransmitting side

When an RLC PDU is transmitted, all its SDUs are moved to the Retransmission (ReTx) queue in case the receiver requests this RLC PDU to be re-transmitted. An SDU can be retransmitted if it falls within the window [VT(A) VT(S)[. In addition, each SDU can be

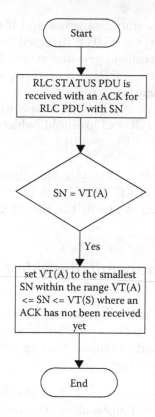

Figure 5.13: RLC AM receiver behavior when an RLC STATUS PDU is received.

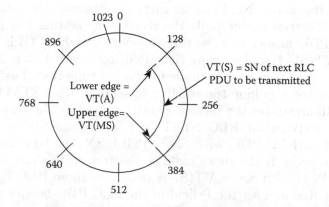

Figure 5.14: RLC AM transmitter sliding window for 10 bits SN.

retransmitted a limited number of times and if the number of retransmission exceeds this limit, the SDU is dropped from the ReTransmission (ReTx) queue and not transmitted any more. It is up to upper layer (TCP or IP) to detect such a PDU loss and requests the retransmission of the TCP/IP packet again.

The maximum number of RLC SDU retransmissions is controlled by the state variable, maxRetxThreshold, which is configured by RRC sublayer as in Table 5.1. Figure 5.15 illustrates the flow chart for retransmission.

Depending on the uplink transmission opportunity size, the transmitter can transmit the RLC PDU as it is or it can segment it into multiple RLC PDU segments to fit the size of the uplink transmission opportunity.

5.5.6   RLC AM transmitting side for polling RLC STATUS PDU

The transmit side can poll the receiver so that the latter can send an RLC STATUS PDU to the transmitter about which RLC PDU is being ACKed or NACKed. Transmitter can poll the receiver if any of the following conditions is met:

■ If either the Transmission (Tx) or the ReTransmission (ReTx) queues will become empty after transmitting an RLC PDU.

■ If the transmit window will be stalled after transmitting an RLC PDU and no more RLC PDU transmission is possible.

If any of the above condition is met, the transmitter can poll the receiver. The transmitter polls the receiver by setting the P field in the RLC PDU header to 1, setting the state variable POLL_SN to be equal to $VT(S)-1$, and starting the t-PollRetransmit timer. The timer t-pollRetransmit acts as a periodic interval of time that if expired, the transmitter keeps polling the receiver to send an RLC STATUS PDU. Figure 5.16 illustrates the polling algorithm used by the transmitter.

Upon receiving an RLC STATUS PDU containing an ACK or NACK for an RLC PDU with SN = POLL_SN, the t-PollRetransmit timer is stopped. If the timer expires, the transmitter can retransmit the RLC PDU with SN = $VT(S)-1$ or any of those RLC PDUs that are not ACKed and set the P field in the RLC PDU header to 1.

Figures 5.17–5.19 show the RLC PDU format that encapsulates either a single RLC SDU, multiple RLC SDUs, or a segment of an RLC SDU, respectivly. The SN field can be 10 or 16 bits in length as configured by the RRC sublayer. Each RLC PDU consists of the RLC

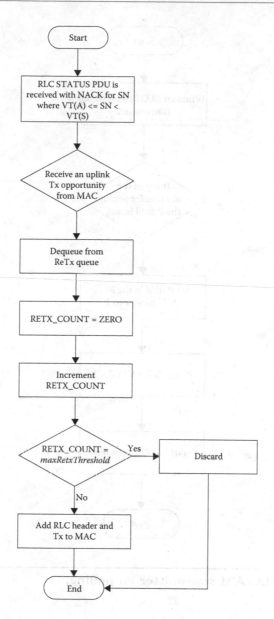

Figure 5.15: RLC AM transmitter on retransmission.

header and payload. The payload can be a single RLC SDU, concatenated RLC SDUs, or a segment of RLC SDU. The header fields are explained and summarized in Table 5.5.

Figure 5.17 shows an RLC PDU where it can be used for a single RLC SDU or a segment of an RLC SDU. If used for a segment of RLC

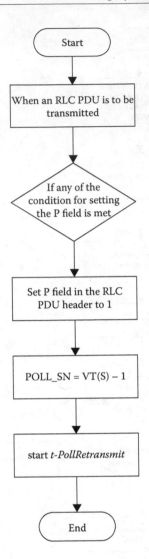

Figure 5.16: RLC AM transmitter on polling.

SDU, the FI (Framing Indicator) field indicates whether this segment is at the beginning of its RLC SDU, middle, or the end of its RLC SDU. FI field has the meaning and format as shown in Table 5.5.

Figure 5.18 shows an RLC PDU format when there are multiple RLC SDUs concatenated in the same RLC PDU. The RLC header has an extension part as there are more than one RLC SDU included in the payload (represented by more than one Payload field). If multiple

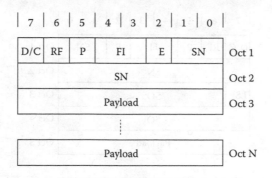

Figure 5.17: RLC PDU for a single RLC SDU or an RLC SDU segment (10 bits SN).

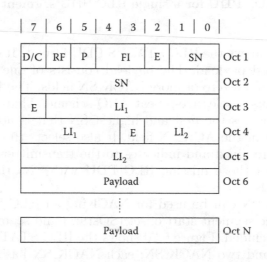

Figure 5.18: RLC PDU for multiple RLC SDUs (10 bits SN).

Payload fields exists, the E and LI fields are present for every Payload field element except the last Payload field.

Figure 5.19 shows an RLC PDU format when there is a single segment of RLC SDU. The RLC payload consists of the RLC SDU segment. This format is also used if a single RLC PDU segment is to be re-segmented again into multiple segments for a retransmission. Segment Offset (SO) indicates the offset of the payload from its original RLC SDU. SO along with the length of the RLC PDU can be used to unambiguously indicate the starting byte and ending byte offset of this segment from the original RLC SDU.

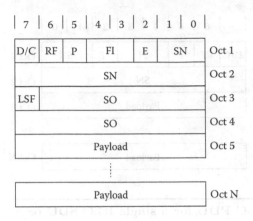

7	6	5	4	3	2	1	0	

Figure 5.19: RLC PDU for a single RLC SDU segment (10 bits SN).

Figure 5.20 shows the RLC STATUS PDU format. It consists of the RLC header and payload. The payload consists of one and only one ACK_SN field and a zero or more NACK_SN fields. The RLC STATUS PDU implements a selective-repeat ARQ scheme. That is, RLC STATUS PDU indicates the first and highest SN that is not yet received by the receiver in the ACK_SN field. It also selects those RLC PDUs that have not received and indicates to the transmitter to repeat the transmissions of those missing RLC PDU with SNs that are in the NACK_SN fields.

The NACK_SN can be used for NACKing an RLC SDU segment. In this case, the segment start offset (SOstart) and segment end offset (SOend) are included. Figure 5.20 shows the RLC STATUS PDU with one ACK_SN and two NACK_SN; each NACK_SN has its own set of SOstart and SOend. Padding can be added to the RLC PDU to keep it octet-aligned.

5.5.7 RLC AM receive side

An RLC AM entity at the receiver side uses a receiving window of size AM window size. A received RLC PDU with an SN must fall in receiving window, [VR(R) VR(MR)[, to be stored in the reception (Rx) buffer, otherwise it is discarded. The AM window is a sliding window protocol used to process received RLC PDUs which can be out-of-order or duplicates.

Table 5.4 shows the state variables used at RLC AM entity acting as a receiver. VR(R) and VR(MR) represent the lower and upper edges of the receiving window. t-Reordering timer can have a value of zero.

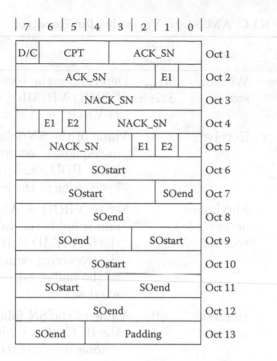

7	6	5	4	3	2	1	0	
D/C	CPT			ACK_SN				Oct 1
ACK_SN							E1	Oct 2
NACK_SN								Oct 3
E1	E2		NACK_SN					Oct 4
NACK_SN					E1	E2		Oct 5
SOstart								Oct 6
SOstart					SOend			Oct 7
SOend								Oct 8
SOend				SOstart				Oct 9
SOstart								Oct 10
SOstart				SOend				Oct 11
SOend								Oct 12
SOend			Padding					Oct 13

Figure 5.20: RLC STATUS PDU (10 bits SN).

A t-Reordering timer with an interval of zero means that upon storing an RLC PDU in the reception (Rx) buffer, RLC can start reassembling the buffered RLC PDUs, reordering them in ascending order of their SNs, stripping the RLC PDU headers, and delivering them to PDCP sublayer.

The receiving window and the state variables can be illustrated as in Figure 5.21. The receiving window can slide indicating the expected range of SNs to be received. VR(MS) stores the first and highest SN that is still missing at the receiver. VR(MS) is used to set the ACK_SN field in the RLC STATUS PDU. VR(H) is used to hold the SN following the highest SN received. When transmitting an RLC STATUS PDU, VR(MS) is set to the first missing RLC PDU SN greater than VR(H).

If the received RLC PDU has its SN in the window [VR(R) VR(MR)[, it is stored in the reception (Rx) buffer. Flow chart for updating the state variables upon reception of an RLC PDU is shown in Figure 5.22. In this figure, it is important to notice that upon storing the received RLC PDU in the reception (Rx) buffer, state variables, VR(H), VR(X), VR(MS), and the receiving window [VR(R) VR(MR)[are updated.

Table 5.4   RLC AM State Variables for Receiver

State Variable	Usage	Initial Value	Meaning
AM_Window_Size	Window size	512, 32768	Determines the receive window size [VR(R) VR(MR)[where VR(MR) = VR(R) + AM window size
VR(R)	Receive	0	Value of the SN following the last in-sequence completely received RLC PDU, and it serves as the lower edge of the receiving window
VR(MR)	Maximum acceptable receive	0	Set to VR(R) + AM_Window_Size, and it holds the value of the SN of the first AMD PDU that is beyond the receiving window and serves as the higher edge of the receiving window
VR(H)	Highest received	0	Value of the SN following the SN of the RLC PDU with the highest SN among received RLC data PDUs
VR(MS)	Maximum STATUS transmit	0	Holds the highest possible value of the SN which can be indicated by ACK_SN when a STATUS PDU needs to be send from Rx to Tx
VR(X)	t-Reordering	0	Holds the value of the SN following the SN of the RLC data PDU which triggered t-Reordering
t-Reordering	Timer	0	A timer value that when expired triggers reordering of received RLC PDU and delivering them to PDCP

Figure 5.22 explains the RLC AM receiver behavior. A received RLC PDU, with SN = x, that has not been received before is stored in the reception (Rx) buffer. Upon storing the RLC in the reception (Rx) buffer, the state variables are first updated according to the following order:

- if $x >$ VR(H), update VR(H) to $x + 1$.

- if $x =$ VR(MS), update VR(MS) to the SN of the first RLC PDU, with SN > VR(MS) which has not been received yet. This also applies to received RLC PDU with SN = x if it is a segment.

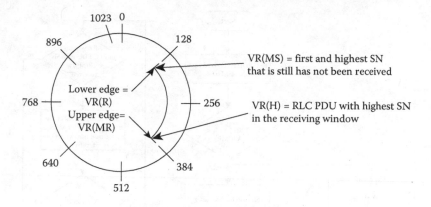

Figure 5.21: RLC AM receiving window.

- if x = VR(R), update VR(R) to the SN of the first RLC PDU, with SN > VR(R) which has not been received yet and update VR(MR) to VR(R)+AM_Window_Size. This also applies to received RLC PDU with SN = x if it is a segment. if VR(R) is updated, this slides the receiving window to a new position [VR(R) V(MR)[.

After the state variables have been updated, and if there is any buffered RLC PDU outside the receiving window, RLC AM entity starts reassembling all buffered RLC PDUs, reordering them, stripping the RLC headers, and delivering them to PDCP sublayer.

5.5.8 RLC STATUS PDU transmission by receive side

The receive side are due to generate and send an RLC STATUS PDU to the transmitter to inform the latter which RLC PDU is ACKed or NACKed. A receiver transmits RLC STATUS PDU to the transmitter either directly or after an amount of time if any of the following condition is met:

- An RLC PDU with P field in its header is set to 1 and it falls outside the receiving window. If the RLC PDU falls in the receiving window, RLC STATUS PDU transmission is delayed until RLC PDU falls outside the receiving window.

- An RLC PDU is detected to be failed reception or if indicated by MAC sublayer. This is only applicable if the enableStatusReportSN-Gap flag is enabled as in Table 5.1.

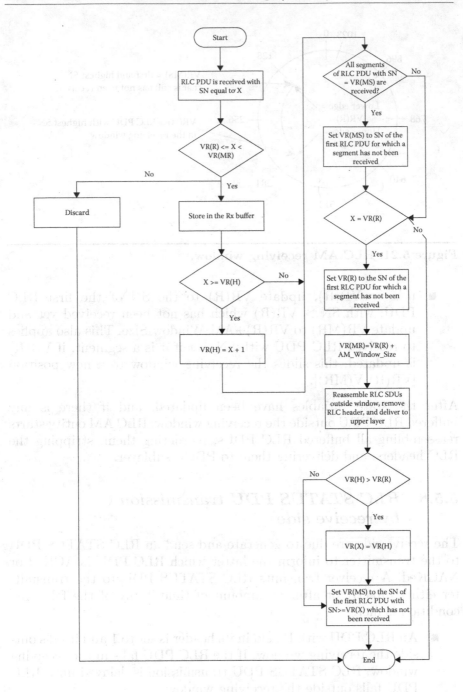

Figure 5.22: RLC AM receiver side.

■ After a reception of an RLC PDU that creates a gap in the SN expected to be received (i.e., indicates a missing RLC PDU). This condition occurs if VR(H) > VR(R).

When an RLC STATUS PDU is to be transmitted by the receiver, it includes an ACK and a zero or more NACK, as shown in Figure 5.20, for all RLC PDUs that have not yet received in the receiving window [VR(R) VR(MS)[.

5.5.9 RLC SDU discard

The PDCP sublayer can request to discard an PDCP PDU or its RLC SDU. In such a case, the RLC can silently discard this RLC SDU if the SDU nor a segment of it has been transmitted yet to the receiver.

5.5.10 Example of an RLC AM transmitter and receiver

To clarify the AM transmitter and receiver operations by an example, consider the example in Figure 5.23. In this example, the RLC SN field is 10 bits in length and the AM_Window_Size is 512 which is half the SN space. The transmitter has a receiving window size of 512, its transmit window, [VT(A) VT(MS)[, is set to [0 512[, and the VT(S) initial value is zero.

Transmitter receives seven RLC SDUs from the upper PDCP sublayer, queues them into the Transmission (Tx) queue. Transmitter receives equal-size uplink transmission opportunities from the MAC subalyer where each opportunity can accommodate a single RLC SDU at a time (including the RLC header). To transmit the seven RLC PDUs, with SNs numbered from 0 to 6, the transmitter assigns an SN equal to VT(S) to each transmitted RLC PDU, increments VT(S), and passes each RLC PDU to the MAC sublayer for transmission to the receiver. Since VT(S) is incremented after each transmission, its value eventually becomes equal to 7.

The receiver also has an AM_Window_Size of 512 and its receiving window, [VR(R) VR(MR)[, is set to [0 512[. At the receiver, state variables, VR(R), VR(H), VR(MS), and VR(X) all have an initial value of zero. VR(MR) has an initial value of VR(R)+AM_Window_Size = 512. The receiver received only RLC PDUs with SN #0, #2, #4, #5 while those RLC PDUs with SN #1, #3, and #6 are lost over the air and have not been received at the receiver.

Since RLC PDU with SN equal to 0, 2, 4, 5 are all within the receiving window, they are all accepted and stored in the reception (Rx)

Table 5.5   RLC PDU Header Fields for UM and AM

Field	Value	Meaning
D/C	0,1	Indicates whether the RLC PDU is an RLC data PDU or RLC control PDU (e.g., STATUS PDU)
RF	0,1	Resegmentation Flag. Indicates whether the RLC PDU contains a single RLC SDU or a segment of an RLC SDU
P	0,1	Polling bit. If set, indicates that the transmitter requests an RLC STATUS PDU report from the receiver
FI	00, 01, 10, 11	Framinf Info. Indicates whether an RLC SDU is segmented at beginning, middle, or end of RLC SDU: 00: segment is exactly the same as the RLC SDU 01: segment is the beginning segment of an RLC SDU 10: segment is the middle segment of an RLC SDU 11: segment is the end segment of an RLC SDU
E	0, 1	0: Data field follows RLC header 1: A set of E field and LI field follows RLC header
SN	[0 31], [0 1023], [0 65535]	Sequence Number. Indicates the SN of the corresponding RLC SDU. For UM, SN is 5 bits. For AM, SN is 10 bits or 16 bits. For an RLC SDU segment, the SN field indicates the SN of the original RLC SDU from which the RLC SDU segment was constructed from
LI	[0 2047], [0 32767]	Length in bytes of the corresponding Payload field present in the RLC PDU. LI can be 11 bit for RLC UM or 15 bits for RLC AM
LSF	0,1	Last Segment Flag. Indicates whether this is the last segment of an RLC SDU or not
SO	[0 32767], [0 65535]	Segment Offset. Indicates the offset of a segment in bytes from the beginning of the original RLC SDU. SO can be 15 bits or 16 bits
CPT	000, 0001	Control PDU Type indicates the type of the RLC control PDU (RLC STATUS PDU or others)
ACK_SN	[0 1023], [0 65535]	The SN of the next not received RLC PDU which is not reported as missing in the STATUS PDU
E1	0, 1	Extension bit 1. Indicates whether a set of NACK_SN, E1, and E2 follows or not

(Continued)

Table 5.5 (*Continued*) **RLC PDU Header Fields for UM and AM**

Field	Value	Meaning
E2	0, 1	Extension bit 2. Indicates whether a set of SOstart, SOend follows the NACK_SN or not
NACK_SN	[0 1023], [0 65535]	Indicates the SN of the RLC PDU or a segment of RLC SDU that has been detected as lost at the receiver
SOstart	[0 32767], [0 65535]	The first byte of a segment of an RLC SDU with SN = NACK_SN that has been detected as lost at the receiver
SOend	[0 32767], [0 65535]	The last byte of a segment of an RLC SDU with SN = NACK_SN that has been detected as lost at the receiver

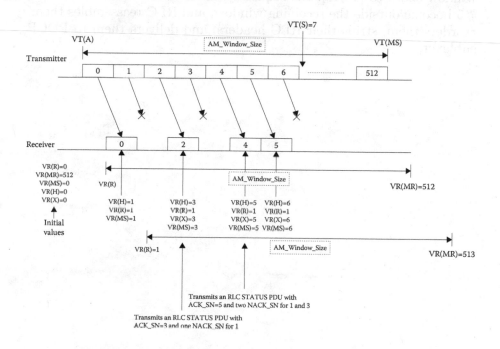

Figure 5.23: RLC AM entity example.

buffer. Each received RLC PDU follows the algorithm in Figure 5.22. After receiving RLC PDU with SN #0, V(R) has slided to a value of 1. After finishing receiving RLC PDU with SN #5, state variables are updated where VR(H) = VR(X) = VR(MS) = 6, and the receiving window, [VR(R) VR(MR)[, is still [1 513[since RLC PDU with SN

#1 is still missing. Since RLC PDU with SN #0 is now outside the receiving window, RLC reassmbles it, reorders it, strips its RLC header, and delivers it to PDCP sublayer.

At the reception of RLC PDU with SN #2 or #4, it triggers a transmissions of an RLC STATUS PDU by the receiver since the reception of either RLC PDU with SN #2 or #4 creates an SN gap where RLC PDU with SN #1 or #3 are missing. When receiving RLC PDU with SN #2, the receiver can transmit an RLC STATUS PDU where ACK_SN is set equal to VR(MS) = 3 and one NACK_SN set equal to 1. At the reception of RLC PDU with SN #4, the receiver transmits an RLC STATUS PDU where ACK_SN is set equal to VR(MS) = 5 and two NACK_SN set equal to 1 and 3.

If the receiver receives RLC PDU with SN #1, #3, the receiving window slides to [6 518[. Consequently, RLC PDUs with SN #2, #4, #5 become outside the receiving window, and RLC reassembles them, reorders them, strips their RLC headers, and delivers them to PDCP sublayer.

Chapter 6

Medium Access Control Sublayer

The MAC sublayer is the lowest sublayer that interfaces directly with the PHY sublayer. Thus, this sublayer performs its functionality in almost real time. The functions of this sublayer are as follows [26]:

- Random access and contention resolution procedures.
- Multiplexes and demultiplex several RLC PDUs to/from a single MAC PDU.
- Hybrid ARQ operation.
- Priority scheduling for signalling and data RBs.
- Mapping of logical channel to/from transport channels.
- Reporting of buffer status, data volume, and scheduling requests.
- Discontinuous reception procedure for conserving battery power.

6.1 MAC Architecture

The MAC architecture is illustrated in Figure 6.1. PDUs transmitted by eNodeB on Broadcast Common Control Channel (BCCH) and Paging Control Channel (PCH) are received by the MAC sublayer and are forwarded transparently to the RLC and then the RRC sublayer

117

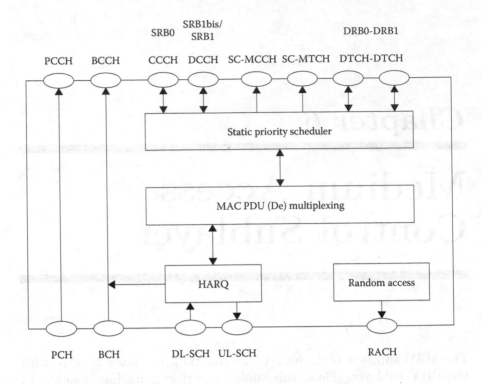

Figure 6.1: MAC architecture for a UE.

without going through the PDCP sublayer. Control- and data-plane PDUs transmitted or received on UL-SCH or DL-SCH, respectively, are processed by the MAC sublayer before they are forwarded to other sublayers. Random access procedure and control are handled by the MAC sublayer and not processed by upper sublayers except to trigger a random access procedure by RRC or to report errors in this procedure to RRC.

SRB0 is mapped to the Common Control Channel (CCCH) which is used and common for all UEs. It is used to transmit or receive RRC PDUs such as RRCConnectionRequest or RRCConnectionReestablishment messages. SRB1 or SRB1bis are mapped to the Dedicated Control Channel (DCCH) which is used and dedicated for a single UE only. DCCH carries RRC PDUs such as RRCConnectionSetupComplete or RRCConnectionRelease PDUs.

Data Radio Bearers (DRBs) are mapped to Dedicated Traffic Channel (DTCH) which is dedicated and used for a single UE only. DTCH carries the data-plane traffic (e.g., TCP/IP) to and from the eNodeB.

When RLC PDUs are received by the MAC sublayer from the RLC sublayer, they are mapped from logical channels (CCCH, DCCH,

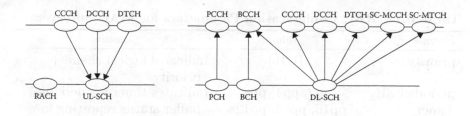

Figure 6.2: Logical and transport channels mapping.

DTCH) to transport channels. After those PDUs are processed by the MAC sublayer, they are passed to the PHY sublayer for transmission using one of the physical channels. If the MAC receives traffic from the PHY channel, it is mapped from a transport channel to one of the logical channels. SC-MCCH and SC-MTCH are used for reception only by UE as they are used for multicast traffic from eNodeB.

Figure 6.2 illustrates how each logical channel is mapped to/from a transport channel. The Random Access Channel (NRACH) has no logical channel since the random access message and procedure are originating and received by the MAC sublayer. BCCH can be received on either the BCH or DL-SCH. BCCH received on BCH is used for receiving MIB-NB PDU whereas BCCH received on DL-SCH is used for other SIBs PDUs.

For UE that only uses control-plane CIoT EPS optimization, as defined in [20], there is only one dedicated logical channel per UE and DTCH is not supported for such a UE.

6.2 RRC Configuration Parameters

RRC sends configuration parameters to MAC to configure random access, SRBs, DRBs, static scheduling, Discontinuous Reception (DRX) configuration, buffer status reporting timers, and scheduling request configurations, as shown in Tables 6.1, 6.2, and 6.3. The MAC configuration parameters are received by RRC from eNodeB during SIB2-NB and SIB22-NB acquisitions, or during the connection establishment procedure explained in Section 3.7.7.

6.3 MAC Procedures

6.3.1 Random access procedure

Random Access (RA) procedure is one of the most important procedures at the MAC sublayer and at the UE since this is the first

Table 6.1   RRC Configuration Parameters for MAC Sublayer

Parameter	Value	Meaning
priority	[1 16]	Indicates logical channel priority
periodicBSR-Timer	pp2, pp4, pp8, pp16, pp64, pp128, infinity	Indicates timer for periodic buffer status reporting in number of NPDCCH periods. pp0 means 0 NPDCCH period
retxBSR-Timer	pp4, pp16, pp64, pp128, pp256, pp512, infinity	Indicates timer for regular buffer status reporting in number of NPDCCH periods
logicalChannel-SR-Prohibit	1 bit	Indicates whether to use the logicalChannelSR-ProhibitTimer timer or not
logicalChannel-SR-ProhibitTimer	pp2, pp8, pp32, pp128, pp512, pp1024, pp2048	Indicates the UE does not transmit Scheduling Request (SR) during the time where timer, logicalChannelSR-ProhibitTimer, is running
drx-Cycle	sf256, sf512, sf1024, sf1536, sf2048, sf3072, sf4096, sf4608, sf6144, sf7680, sf8192, sf9216	Indicates the DRX cycle in number of subframes
drx-StartOffset	[0 255]	Indicates the start offset of DRX cycle in number of sub-frames by step of (drx-cycle/256)
onDuration-Timer	pp1, pp2, pp3, pp4, pp8, pp16, pp32	Indicates the onDuration period of a DRX cycle
drx-Inactivity-Timer	pp0, pp1, pp2, pp3, pp4, pp8, pp16, pp32	Indicates InactivityTimer in number of NPDCCH periods
drx-Retransmission-Timer	pp0, pp1, pp2, pp4, pp6, pp8, pp16, pp24,pp33	Indicates RetransmissionTimer in number of NPDCCH periods
drx-UL-Retransmission-Timer	pp0, pp1, pp2, pp4, pp6, pp8, pp16, pp24, pp33, pp40, pp64, pp80, pp96, pp112, pp128, pp160, pp320	Indicates ULRetransmission-Timer in number of NPDCCH periods

Table 6.2   Random Access Parameters Provided by RRC Sublayer

Parameter	Value	Meaning
nprach-ParametersList	List	A list of entries where each entry as in Table 6.3 contains an NRACH resource available on anchor or non-anchor subcarrier
ul-ConfigList	List	A list of entries where each entry contains parameters for NRACH resources on a non-anchor carrier
ra-Response-WindowSize	pp2, pp3, pp4, pp5, pp6, pp7, pp8, pp10	Indicates duration of the RA response window in NPDCCH period (e.g., pp2 indicates 2 NPDCCH periods)
mac-Contention-ResolutionTimer	pp1, pp2, pp3, pp4, pp8, pp16, pp32, pp64	Indicates Timer value for contention resolution
ra-CFRA-Config	True	If presents, indicates the activation of Contention-Free Random Access (CFRA)
rsrp-ThresholdsPrach-InfoList	List	A list of two RSRP threshold levels
nprach-Probability-AnchorList	List	A list of selection probability for each NRACH resource on anchor carrier
nprach-Probability-Anchor	zero, oneSixteenth, oneFifteenth, oneFourteenth, oneThirteenth, oneTwelfth, oneEleventh, oneTenth, oneNinth, oneEighth, oneSeventh, oneSixth, oneFifth, oneFourth, oneThird, oneHalf	Indicates selection probability for an NPRACH resource on the anchor carrier

(Continued)

Table 6.2 (*Continued*)   Random Access Parameters Provided by RRC Sublayer

Parameter	Value	Meaning
preamble-TransMax-CE	n3, n4, n5, n6, n7,n8, n10, n20, n50, n100, n200	Indicates the maximum number of preamble transmissions
nprach-CP-Length-r13	us66dot7, us266dot7	Indicates whether to use preamble format 0 or format 1 as in Table 7.32

Table 6.3   Parameters of an NRACH Resource on Anchor or Non-Anchor Carrier

Parameter	Value	Meaning
nprach-Periodicity	ms40, ms80, ms160, ms240, ms320, ms640, ms1280, ms2560	Determines the radio frame where an NRACH resource is available
nprach-StartTime	ms8, ms16, ms32, ms64, ms128, ms256, ms512, ms1024	Determines the start time after the start of an uplink radio frame in multiple of 1 ms from the start of the radio frame
nprach-SubcarrierOffset	n0, n12, n24, n36, n2, n18, n34	Determines the first subcarrier of an NRACH resource
nprach-NumSubcarriers	n12, n24, n36, n48	Total number of subcarriers available for an NRACH resource in a radio frame
numRepetitions-PerPreamble-Attemp	n1, n2, n4, n8, n16, n32, n64, n128	Indicates the number of random access preamble repetitions per attempt for each NPRACH resource
nprach-SubcarrierMSG3-RangeStart	zero, oneThird, twoThird, one	A fraction for calculating two groups of subcarriers that one of them can be used for indication of multi-tone support
nprach-NumCBRA-StartSubcarriers	n8, n10, n11, n12, n20, n22, n23, n24, n32, n34, n35, n36, n40, n44, n46, n48	Determines the number of subcarriers where the UE can randomly select a start subcarrier to transmit RAP

procedure a UE initiates to connect to an eNodeB. The main purpose of RA procedure is to achieve uplink synchronization and obtain an uplink grant for starting the RRC connection establishment and NAS Attach procedures. The RA procedure consists of a sequence of four messages, Msg1, Msg2, Msg3, and Msg4. Msg1 is the RA Preamble, Msg2 is the Random Access Response (RAR), Msg3 is the RRC PDU transmitted in the UL grant, and Msg4 is the message received from the eNodeB (i.e., Contention Resolution Identity). Basic information about the RA procedure is broadcasted to UE in SIB2-NB.

A UE that moves from IDLE to CONNECTED mode needs to trigger the random access procedure. An eNodeB that needs to connect to a UE in IDLE mode can send a message on the downlink NPDCCH ordering the UE to initiate an RA procedure. The RA procedure can be contention-based or contention-free procedure, can be performed on anchor or non-anchor carrier, and consists of two sub-procedures:

- Random Access

- Contention Resolution

The RA procedure can be initiated by the MAC sublayer itself, the RRC sublayer, or by an order from eNodeB through NPDCCH. If initiated by the MAC sublayer, the RA procedure is contention-based. If the eNodeB orders the UE to initiate the RA procedure, then it can be contention-free procedure.

6.3.2 Random access exchange

The purpose of the RA is to send a Random Access Preamble (RAP) to the eNodeB that can enable the UE in IDLE mode to transmit an RRCConnectionRequest PDU, perform RRC connection establishment procedure explained in Section 3.7.7, and finally move from IDLE to CONNECTED mode. Table 6.2 summarizes the random access procedure parameters provided by the RRC sublayer to the MAC sublayer.

eNodeB allocates a number of NRACH resources for random access preamble transmissions. An NRACH resource is identified by a number of subcarriers in frequency domain and a specific start time in uplink radio frame. Each anchor carrier has a maximum of three NRACH resources and each non-anchor carrier has a maximum of three NRACH resources as well.

Each cell is partitioned into one to three coverage enhancement levels. Coverage enhancement levels are used to accommodate devices with different level of channel quality to eNodeB (i.e., good or poor channel quality). UE determines its coverage enhancement level based on its measured reference signal received power (RSRP)[1]. Each coverage enhancement level is mapped to a single NRACH resource on anchor carrier and zero or one NRACH resource on each non-anchor carrier. The UE selects one of the NRACH resource based on its coverage enhancement level. As the same coverage enhancement level can have one or more NRACH resources (on anchor and non-anchor carriers), the UE can select one of them according to a probability distribution.

The number of enhanced coverage levels is equal to one plus the number of RSRP thresholds in rsrp-ThresholdsPrachInfoList. Enhanced coverage levels are numbered from 0 to 2 and they map to an NPRACH resource in an increasing, numRepetitionsPerPreambleAttempt, order for each NRACH resource [27]. NB-IoT device with best channel quality to eNodeB selects enhanced coverage level #0 whereas NB-IoT device with poorest channel quality to eNodeB selects enhanced coverage level #2.

Concurrent channel access from massive number of NB-IoT devices can congest the RA channel. For this reason, more NRACH resources are provided on anchor or non-anchor carriers, or by ordering the UE to perform contention-free access.

RRC configures the MAC sublayer with a number of NRACH parameters that are used to select the NRACH resource in the uplink slots to transmit the random access preamble. The PHY uses a single starting subcarrier with additional three frequency-hopping subcarriers for transmission of the random access preamble. RRC provides the following parameters to the MAC sublayer:

- The number of NRACH resources that can be used for transmission of a random access preamble. A maximum of three NRACH resources are available on anchor carrier and each non-anchor carrier. The available set of NRACH resources on the anchor carrier are indicated by nprach-ParametersList (in SIB2-NB), and those resources on the non-anchor carriers are indicated by ul-ConfigList (in SIB22-NB).

- RA response Windows size, ra-ResponseWindowSize.

- Contention Timer, mac-ContentionResolutionTimer.

- Whether the UE to activate contention-free or contention-based random access parameter as indicated by ra-CFRA-Config.

UE first selects its coverage enhancement and its mapped NRACH resource according to the measured RSRP. UE, then, selects the random

---

[1]RSRP measurement is performed on the anchor carrier.

access preamble in the NRACH resource for transmission. That is, once the random access preamble and NRACH resource are determined, the UE transmits the RAP[2]. Preamble collision occurs if two or more UEs choose the same random access preamble and the same NRACH resource. Collision can be avoided if the eNodeB orders UE to use an explicit random access preamble and an NRACH resource.

If the UE has more than one NRACH resource available for its coverage enhancement level on anchor and non-anchor carriers, a single NRACH resource is selected on either anchor or non-anchor carrier. The selection probability for an NRACH resource on anchor carrier is equal to nprach-ProbabilityAnchor. There can be zero or more NRACH resources on each of the non-anchor carrier. The probability of selecting each NRACH resource on non-anchor carrier is (1-nprach-ProbabilityAnchor)/(number of non-anchor NPRACH resources). nprach-ProbabilityAnchor is provided in the corresponding entry in nprach-ProbabilityAnchorList (in SIB22-NB). Typically, if the random access procedure is initiated by an NPDCCH order, the carrier is explicitly chosen through Carrier Indication field (as explained in Section 7.10.9.13).

Table 6.3 indicates the RRC parameters to identify the time and frequency location of an NRACH resource. A radio frame, n_f, contains an NRACH resource if n_f mod (nprach-Periodicity/10) = 0. Within this radio frame, the NRACH resource starts nprach-StartTime after the start of this radio frame.

In Table 6.3, an NRACH resource consists of a number of subcarriers, nprach-NumSubcarriers. A subcarrier is identified by the subcarrier index in the range:

nprach-SubcarrierOffset + [0 nprach-NumSubcarriers − 1].

Table 6.3 also includes the RRC parameters if the UE supports multitone transmission for Msg3. Each NRACH resource contains a set of, nprach-NumSubcarriers, subcarriers which can be partitioned into one or two groups for single or multi-tone Msg3 transmission. Each group is referred to as a random access preamble group. UE selects only a single group and a random access preamble within the group.

Group A is defined by:

nprach-SubcarrierOffset + [0 (nprach-NumCBRA-StartSubcarriers
    × nprach-SubcarrierMSG3-RangeStart) − 1],

---

[2]A random access preamble is also called "Signature."

and Group B is given by:

$$\text{nprach-SubcarrierOffset}$$
$$+ [(\text{nprach-NumCBRA-StartSubcarriers}$$
$$\times \text{ nprach-SubcarrierMSG3-RangeStart})$$
$$\text{nprach-NumCBRA-StartSubcarriers} - 1],$$

where each subcarrier of a random access preamble group corresponds to a random access preamble.

If a subcarrier index is chosen by the UE; either from Group A or Group B (in case of multi-tone), additional three subcarrier indices are selected in a frequency-hopping manner. The four subcarrier indices are co-located within a block of 12 subcarriers.

The RA procedure can start by a trigger from the UE itself (either by the MAC sublayer itself or RRC sublayer itself). In this case, the UE selects a random access preamble which can collide with the transmissions of other UEs (i.e., contention-based).

Random access procedure can also be triggered by an order from the eNodeB on the NPDCCH. For example, when eNodeB receives DL data for this UE and it needs the UE to move to CONNECTED mode, eNodeB transmits an NPDCCH order. If the eNodeB transmits an NPDCCH order, it can indicate an explicit and assigned random access preamble to be used by the UE. In this case, collision can be avoided by the UEs (i.e., contention-free [2]). NB-IoT UE category NB1 supports only contention-based random access transmission on anchor carrier and on non-anchor carrier [27].

If a subcarrier indication field is signalled through NPDCCH order[3], then the parameter, ra-PreambleIndex, is set equal to the signalled subcarrier indication field. Both ra-PreambleIndex and the chosen NRACH resource are considered to be explicitly signalled. If ra-PreambleIndex is not zero and ra-CFRA-Config is enabled, the MAC sublayer chooses the random access preamble[4] according to the following:

$$\text{nprach-SubcarrierOffset}$$
$$+ \text{nprach-NumCBRA-StartSubcarriers}$$
$$+ (\text{ra-PreambleIndex mod}$$
$$(\text{nprach-NumSubcarriers} - \text{nprach-NumCBRA-StartSubcarriers})).$$

---

[3]Subcarrier indication field is explained in Section 7.10.9.13.
[4]Random access preamble is the same as starting subcarrier index.

Otherwise, if ra-CFRA-Config is disabled, the random access preamble is chosen according to the following:

nprach-SubcarrierOffset

+ (ra-PreambleIndex mod nprach-NumSubcarriers).

However, If the ra-PreambleIndex is zero while an NRACH resource is chosen (that selects one of the two multi-tone groups), UE selects the random access preamble randomly from that group. If only one group exists, the UE selects the random access preamble randomly from that group.

If the UE triggers the RA procedure by itself (i.e., through MAC or RRC sublayers), the UE selects one of NRACH resources on the anchor or non-anchor carriers according to the enhanced coverage level and their probability distribution, nprach-ProbabilityAnchor. After selecting an NRACH resource, the UE can either select one of the two groups and select a random access preamble randomly within that group.

Once the NRACH resource and the random access preamble are selected, the MAC passes them to the PHY and inform it to transmit an RAP to the eNodeB repeated a number of, numRepetitionsPer-PreambleAttempt, times.

If the UE transmits the RAP, it keeps watching out for the response in the number of downlink frames for a window of frames after the end of the RA transmission. The response window starts at the subframe that contains the end of the last preamble repetition plus 41 subframes (if number of preamble repetition is 64 or more) or 4 subframes (if number of preamble repetition is less than 64) and has length ra-ResponseWindowSize. The UE waits for an RAR by looking for an NPDCCH scrambled with RA-RNTI where RA-RNTI is:

$$\text{RA-RNTI} = 1 + \text{floor}(SFN_{id}/4) + 256 \times carrier_{id},$$

where SFN_{id} is the index of the first radio frame of the NPRACH resource and $carrier_{id}$ is the index of the UL frequency carrier ID. For the anchor carrier, $carrier_{id}$ is 0. The RAR format that is expected to be received by the UE is illustrated in Figure 6.3. If the UE receives an RAR in the response window scrambled with RA-RNTI, the RAR can be designated for this UE. If the RAR is designated for this UE, the RAR contains UL grant indicating an available UL grant for the UE to transmit its Msg3. The RAR also comes with a Temporary C-RNTI which is used to encode all transmissions on NPDCCH. UE decodes the NPDCCH using this temporary C-RNTI. The content of the UL grant is interpreted as in Table 6.4.

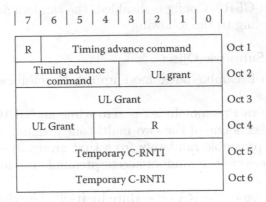

7	6	5	4	3	2	1	0

R	Timing advance command	Oct 1
Timing advance command	UL grant	Oct 2
UL Grant	Oct 3
UL Grant	R	Oct 4
Temporary C-RNTI	Oct 5
Temporary C-RNTI	Oct 6

Figure 6.3: Msg2: Random access response (RAR) format.

Table 6.4   UL Grant Content

Information	Size (Bits)	Meaning
Subcarrier spacing (Δf)	1	If set to 0, $\Delta f = 3.75$ KHz otherwise if 1, $\Delta f = 15$ KHz
Subcarrier indication (I_{sc})	6	Determines the UL subcarriers allocated, n_{sc}
Scheduling delay (I_{Delay})	2	The number of subframes elapsed from the end of NPDCCH until the first UL slot of the NPUSCH
Modulation and coding scheme (I_{MCS})	3	Determines the modulation scheme, transport block size, and number of uplink resource unit for Msg3 transmission as shown in Table 6.5
Msg3 Repetition number (N_{Rep})	3	Determines the number of repetition for Msg3

Table 6.5   Modulation, Number of Resource Units, and Transport Block Size for MCS Index

MCS Index (I_{MCS})	Modulation $\Delta f = 3.75/15$ KHz, $I_{sc} = 0 - 11$	Modulation $\Delta f = 15$ KHz, $I_{sc} > 11$	Number of RUs (N_{RU})	Transport Block Size (Bits)
000	$\pi/2$ BPSK	QPSK	4	88
001	$\pi/4$ QPSK	QPSK	3	88
010	$\pi/4$ QPSK	QPSK	1	88

If the UE successfully received its designated RAR, it concludes that the RA procedure is completed successfully and starts the contention resolution procedure.

A number of random access responses (RAR) can be received by the UE in a single MAC PDU that is not designated to the same UE but other UEs as shown in Figure 6.4. The UE can identify whether this response is designated for itself or not by matching the Random Access Preamble Identifiers (RARID) received with the transmitted random access preamble[5]. If they match, it means the response is designated to this UE and this indicates the RA procedure is completed successfully.

If no RAR is received during the RA Window size, ra-Response-WindowSize, the UE starts a back-off procedure where it selects a back-off value that is uniformly-distributed in the range [0 Back-off] and re-transmits a new RAP. The UE can re-run the random access procedure up to a maximum number of times, preambleTransMax-CE, as set by the RRC sublayer.

An RAR can contain a back-off index message that indicates the back-off value as shown in Table 6.6. If no back-off index message is received, the UE sets its back-off value to zero.

6.3.3 Contention resolution

When the UE completes the RA procedure successfully, it starts the contention resolution mechanism.

The RAR received by the UE is as shown in Figure 6.3. Table 6.5 shows the UL specification included in the RAR message. UE uses these parameters to start its Msg3 transmission. Typically, a Msg3 is

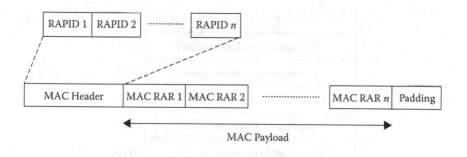

Figure 6.4: Number of random access responses (RARs) received in one MAC PDU.

---

[5]The RAR ID is the starting subcarrier index selected when the random access procedure is triggered.

Table 6.6 Back-off Parameter Values

Index	Back-Off Parameter Value (ms)
0	0
1	256
2	512
3	1024
5	4096
6	8192
7	16384
8	32768
9	65536
10	131072
11	262144
12	524288

an RRCConnectionRequest PDU so that the UE moves from IDLE mode to CONNECTED mode.

When the UE transmits or re-transmits the Msg3 in the UL grant, it starts or restarts the mac-ContentionResolutionTimer timer. During the running of this timer, the UE looks for an NPDCCH transmission addressed to its Temporary C-RNTI.

If an NPDCCH transmission is addressed to the UE temporary C-RNTI is received by the UE, the UE decodes the corresponding MAC PDU received on DL-SCH. If the MAC PDU contains a contention resolution identity MAC control element, as in Figure 6.5, that matches the transmitted Msg3, the UE stops the mac-ContentionResolutionTimer

	7	6	5	4	3	2	1	0	

UE contention resolution identity	Oct 1
UE contention resolution identity	Oct 2
UE contention resolution identity	Oct 3
UE contention resolution identity	Oct 4
UE contention resolution identity	Oct 5
UE contention resolution identity	Oct 6

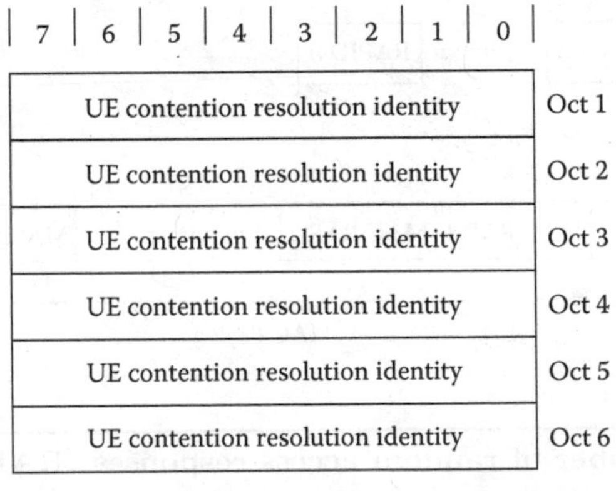

Figure 6.5: Msg4: contention resolution identity MAC control element.

timer and considers the contention resolution procedure as completed successfully. Upon successful completion of the contention resolution procedure, the UE sets its C-RNTI to the temporary C-RNTI and discards the temporary C-RNTI.

If the mac-ContentionResolutionTimer timer expired, the MAC PDU can not be decoded successfully, or the MAC PDU does not contain a contention resolution identity that matches the transmitted Msg3, the contention resolution procedure is considered as not successful. If the contention resolution procedure is not successful, the UE can decide to repeat the RA procedure all over again.

Figure 6.6 illustartes the random access and contention resolution procedures. Initialliy, the UE is in IDLE mode and is camping on a cell. When the UE or the RRC decides to connect to the eNodeB,

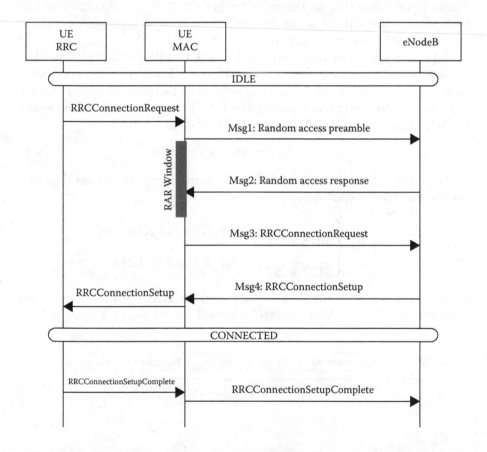

Figure 6.6: Exchange of Msg1, Msg2, Msg3, and Msg4 between UE and eNodeB.

MAC transmits Msg1 (random access preamble). Upon receiving Msg2 (random access response) which contains the UL grant, as in Table 6.4, the UE transmits Msg3. MAC uses the UL grant to transmit Msg3 (RRCConnectionRequest). The MAC expects to receive Msg4 which can contain both an RRCConnectionSetup and MAC control element containing matched contention resolution identity. At this point, the UE can move to the CONNECTED mode and transmits the RRC-ConnectionSetupComplete PDU.

6.3.4 Timing advance command

In the RAR, the UE receives an 11-bit timing advance (TA) command. The UE uses this command to adjust the timing of downlink and uplink frames relative to each other. The timing of a downlink and uplink frame is as illustrated in Figure 6.7 where $0 \leq N_{TA} \leq 20512$. The UE uses the anchor carrier as timing reference irrespective whether there is a non-anchor carrier configured or not.

Upon reception of this command, the UE adjusts the transmission timing of the uplink channel; NPUSCH and NPRACH. This command indicates to the UE that a change is needed to be done for uplink timing with regard to the current timing the UE is currently using. The exact timing alignment value, N_{TA}, is calculated from

$$N_{TA} = 16 \times T_A, \tag{6.1}$$

where $T_A = 0, 1, 2, \ldots, 1282$ and T_s is the sampling interval that depends on the subcarrier spacing:

$$T_s = \begin{cases} \dfrac{1}{2048 \times \Delta f} & \text{for } \Delta f = 15 \text{ KHz} \\ \dfrac{1}{8192 \times \Delta f} & \text{for } \Delta f = 3.75 \text{ KHz.} \end{cases}$$

In addition to receiving a TA command in RAR, eNodeB can also transmit a TA as a MAC control element as shown in Figure 6.8. It

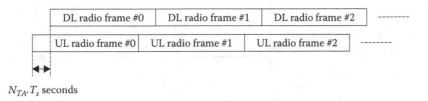

$N_{TA}.T_s$ seconds

Figure 6.7: Downlink and uplink timing.

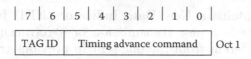

Figure 6.8: TA command as MAC control element.

consists of TAG identity which is set to zero and a 6-bit TA command value, $T_A \in \{0, 1, \ldots, 63\}$. The TA command value is used to adjust old timing, $N_{TA,old}$, to a new timing $N_{TA,new}$, by the value of T_A received according to:

$$N_{TA,new} = N_{TA,old} + (T_A - 31) \times 16. \tag{6.2}$$

Depending on the value of TA, the new timing can be advanced or delayed relative to the current timing.

If the TA command is received in DL frame, n, the UE start applying the new TA from the first available uplink NPUSCH slot following the $(n + 12)$th DL subframe.

6.4 Data Transfer

6.4.1 Downlink data reception

Downlink (DL) data transmitted to the UE are sent in DL-SCH. UE keeps monitoring the NPDCCH in each subframe[6] to know whether it has data designated for it in the current frame or not. The NPDCCH carries information for HARQ processing. This includes the HARQ process number associated with this subframe.

A UE has only a maximum of two HARQ processes running for DL-SCH. HARQ process for the DL-SCH is a stop-And-Wait process and is asynchronous and adaptive. In a Stop-and-Wait process, packets of a HARQ process are first transmitted then the HARQ process stops to receive an ACK/NACK before transmitting the next packet. While waiting on the ACK/NACK of an HARQ process, packets of a different HARQ process can be transmitted in parallel. Stop-And-Wait HARQ process increases the Round Trip Time (RTT) and decreases throughput.

In asynchronous HARQ, the UE follows what the NPDCCH asks the UE to do (i.e., either to do a transmission or a retransmission) and HARQ process ID is signalled in NPDCCH. Adaptive HARQ means that the same packet can be retransmitted using a different modulation scheme.

---

[6]A subframe duration is also called Transmission Time Interval (TTI).

NPDCCH contains a New Data Indicator (NDI) bit. It tells UE whether this data is a new transmission or a retransmission. If NDI bit is toggled (different from the one sent in previous transmission), it means that a new data is transmitted in downlink for the given HARQ process.

Received MAC PDU is stored in a soft buffer and can be combined with any reception of a retransmission of the same MAC PDU. The UE decodes the combined MAC PDU and depending on whether the decoding is successful or not, it generates either an ACK or NACK to be transmitted to the eNodeB.

When the UE receives a MAC PDU on its DL-SCH, it forwards the HARQ information (HARQ process ID) to its HARQ process along with the received MAC PDU. The MAC PDU can be the first transmission of this MAC PDU or a retransmission of it. The HARQ process combines received MAC PDU with what in the soft buffer (if any) and try to decode the MAC PDU. Upon a successful decoding of the MAC PDU, MAC sublayer continue to disassemble and demultiplex the MAC PDU. Otherwise, if the MAC PDU is not successfully decoded, the UE stores or combines the received MAC PDU with that in the soft buffer. After reception of the MAC PDU, UE either transmits an ACK or NACK indicating whether the MAC PDU is successfully decoded or not.

The UE receives a data unit at the physical sublayer in a form of a transport block. Each downlink transport block is transmitted by eNodeB multiple times equal to a repetition parameter value provided by NPDCCH. This parameter provides the number of subframes where a DL transmission is repeated multiple subframes. For a transport block transmitted to the UE, a number of retransmissions are repeated in multiple subframes as indicated by NPDCCH. The UE transmits a single ACK/NACK for the transport block and its repetitions. NPDCCH downlink assignment, corresponding to a new transmission or a retransmission of transport block, is received after the last repetition of the transport block. Further details about downlink NPDCCH assignment are explained in Section 7.10.9.13.

ACK/NACKs sent on uplink in response to downlink transport block reception are sent on NPUSCH. Retransmissions are scheduled through NPDCCH.

6.4.2 Uplink data transmission

The UE transmits data or control PDUs by encapsulating them in MAC PDU. The UE monitor's the NPDCCH to know about any UL assignment assigned to the UE.

There is a maximum of two HARQ processes for UL-SCH. UL HARQ uses Stop-And-Wait protocol. The HARQ process is responsible for transmission and retransmission of a transport block. The uplink HARQ process is asynchronous and adaptive. In asynchronous adaptive HARQ, the process associated with a subframe is based on the received UL grant except for UL grant in RAR. In other words, in asynchronous adaptive HARQ, the UE follows what the NPDCCH asks the UE to do (i.e., either perform a transmission or a retransmission). For UL transmission with UL grant in RAR, HARQ process identifier 0 is used.

Each HARQ process has a buffer that stores the MAC PDU to be transmitted. The buffer contains another variable; CURRENT_IRV. CURRENT_IRV indicates the index of the current redundancy version (RV) where the RV are defined to be among the sequence $\{0, 2, 3, 1\}$. In asynchronous HARQ operation, UL retransmissions are triggered by adaptive retransmission grants. ACK/NACK for an UL transmission is signalled implicitly by the NDI in the DCI.

Each transmitted transport block on NPUSCH is repeated a number of times equal to a repetition paramater value provided by NPDCCH. An uplink grant corresponding to a new transmission or a retransmission of the initial transmission is only received after the last repetition of the previous transmissions. Further details about uplink NPUSCH transmission are explained in Section 7.10.9.11.

6.5 Discontinuous Reception

DRX is the procedure used by the UE MAC to conserve energy and battery. This procedure works in IDLE and CONNECTED modes. The DRX behavior in IDLE mode is described in Section 3.12. In DRX mode, the UE sleeps where it shutdowns its transceiver and wake up to monitor NPDCCH occasionally.

eNodeB configures the DRX parameters to be used by the UE through the RRC connection establishment or reconfiguration procedures. DRX enables the UE to sleep and wake up only on specific intervals to monitor NPDCCH or to transmit and receive any downlink or uplink messages on NPDSCH or NPUSCH. The DRX parameters, as shown in Table 6.1, are:

- drx-Cycle: Indicates the length of the DRX cycle in subframes which include ON time followed by an OFF (inactivity) time.

- drx-StartOffset: Indicates the DRX offset used to calculate the starting subframe number for DRX cycle.

■ onDurationTimer: Indicates the number of consecutive NPDCCH-subframe(s) at the beginning of each DRX cycle (when the UE is ON). It is the number of subframes over which the UE read NPDCCH during every DRX cycle before entering the power saving mode (UE is OFF).

■ drx-InactivityTimer: Indicates the number of consecutive NPDCCH-subframe(s) for which the UE should be active after successfully decoding an NPDCCH indicating a new transmission (UL or DL). This timer is restarted upon receiving NPDCCH for a new transmission (UL or DL). Upon the expiry of this timer, the UE sleeps.

■ drx-RetransmissionTimer: Indicates the maximum number of consecutive NPDCCH-subframes until a DL retransmission is received. The UE should be monitoring NPDCCH when a retransmission from the eNodeB is expected by the UE.

■ drx-ULRetransmissionTimer: Indicates the maximum number of consecutive NPDCCH-subframe(s) until a grant for UL retransmission is received.

A DRX cycle is synchronized at both UE and eNodeB. Both UE and eNodeB know at what time the UE is asleep or awake. eNodeB can thus schedule downlink traffic to the UE accordingly.

A UE can read NPDCCH only when it is awake otherwise if it is asleep, it does not read NPDCCH. For uplink traffic, it is not affected during wake up or sleep as UE can transmit a scheduling request (SR) when it wakes up to request uplink transmission opportunity. eNodeB can still force the UE into sleep by transmitting DRX MAC control element to the UE.

DRX cycle is started in a subframe that fulfills:

$$[(SFN \times 10) + \text{subframe number}] \mod (\text{drx-Cycle}) = \text{drxStartOffset}.$$

When the DRX cycle starts, the OnDuration timer starts for an interval of onDurationTimer. During the onDurationTimer time interval, UE keeps monitoring NPDCCH for any downlink grant, uplink grant, or retransmission. It can also receive data on NPDSCH or transmit on NPUSCH if there is a downlink assignment or uplink grant. Figure 6.9 shows the DRX timing when there is no NPDCCH designated for the UE.

If during the onDurationTimer time interval, an NPDCCH designated for the UE is received that indicates a downlink or uplink transmission, the UE starts the drx-InactivityTimer timer for a time interval

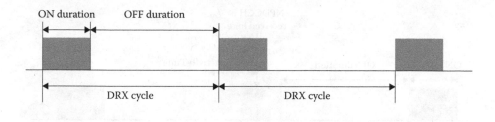

Figure 6.9: DRX procedure when there is no NPDCCH received.

of drx-InactivityTimer to continue to be awake. Figure 6.10 shows the DRX timing when an NPDCCH is received by the UE. The UE wake up time is expanded by the drx-InactivityTimer timer.

If during the onDurationTimer time interval, an NPDCCH designated for the UE is received that indicates MAC PDU that contains DRX MAC control element. The UE initially starts the drx-InactivityTimer but later discovers a DRX MAC control element is received. After the UE process the DRX MAC control elements, it stops both timers; drx-InactivityTimer and onDurationTimer and goes into sleep. This is illustrated in Figure 6.11.

If the UE receives an NPDCCH indicating downlink transmission or UL grant, the UE starts a HARQ RTT timer after receiving the NPDSCH or transmitting the NPUSCH. The HARQ RTT timer is started so that it indicates the minimum amount of subframes before either a DL HARQ retransmission or an UL HARQ retransmission grant is expected. When the HARQ RTT timer is started, UE can go asleep.

If the HARQ RTT timer for DL transmission expired, UE checks if the stored MAC PDU can be decoded successfully or not. If the

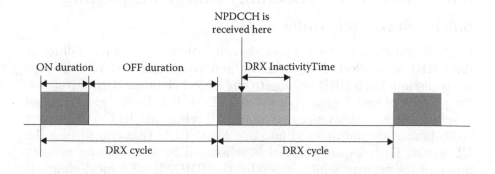

Figure 6.10: DRX procedure when an NPDCCH is received.

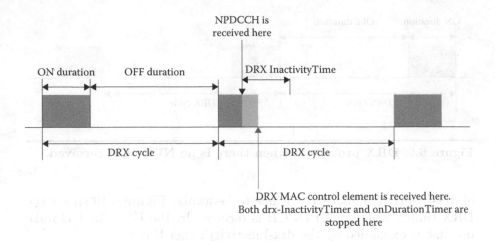

Figure 6.11: DRX procedure when an NPDCCH and DRX MAC control element are received.

stored MAC PDU cannot be decoded successfully, the UE starts drx-InactivityTimer and drx-RetransmissionTimer. If HARQ RTT timer for UL grant expired, UE starts drx-InactivityTimer and drx-ULRetransmissionTimer. By starting any of these timer, the UE wakes up.

The DRX cycle starts only if none of the timers are running; onDurationTimer, drx-InactivityTimer, drx-RetransmissionTimer, drx-ULRetransmissionTimer, or mac-ContentionResolutionTimer, otherwise the DRX cycle is not started.

6.6 MAC PDU Assembly and Multiplexing

6.6.1 Static scheduler

Control and data PDUs queued at each logical channel (signalling or data RB) are collected to assemble and construct a MAC PDU for transmission. Each DRB belongs to one logical channel. RRC advertises the priority of each logical channel as in Table 6.1. Each logical channel is scheduled in a strict priority order. All scheduled RLC PDUs for a radio bearer are multiplexed into one MAC PDU that can fit into the UL grant. Each logical channel is scheduled in a decreasing priority order of its priority which is set by the RRC. If all logical channels have equal priority, then they are scheduled equally (or in a round-robin fashion).

Depending on the size of the UL grant, the MAC sublayer might segment an RLC PDU to fit into the UL grant. On the other hand, if the UL grant is big and there is no more RLC PDUs queued to be scheduled, a Padding MAC control element is added to fill up the UL grant size.

In addition to transmitting and receiving control and data PDUs for the SRB and DRB, respectively, the UE can transmit MAC control elements that are used to exchange control information with the eNodeB. The UE follows a strict priority order for transmitting MAC control elements and data for radio bearers. That is, UE assembles and constructs the MAC PDU by multiplexing the following messages in strict order:

- MAC control element for C-RNTI or data from UL CCCH (SRB0).

- MAC control element for Buffer Status Report (BSR).

- Data from SRB1, SRB1bis, or DRBs. Priority for SRB1 is shown in Table 3.2.

- MAC control element for BSR included for padding.

6.6.2 *Buffer status report*

The UE can transmit a MAC control element, called BSR, to indicate how much data is queued at the UE. This includes data for both SRBs and DRBs. All logical channels belong to one Logical Channel Group (LCG). There are three types of BSR; Regular BSR, Padding BSR and periodic BSR. They are triggered according to the following:

- Regular BSR: This is triggered when an uplink data becomes available for transmission at the RLC or PDCP sublayer for any of the SRB or DRBs for a first time only. It is also triggered when there is no more data available for transmission. If a regular BSR is to be transmitted, a timer, logicalChannelSRProhibitTimer, is started that prohibits the UE for transmitting a request to eNodeB asking for uplink grant. The timer is started only if logicalChannelSR-Prohibit is enabled.

- Padding BSR: This is triggered if the uplink grant is big enough and there is no more RLC PDUs to fill up the uplink grant. The remaining unused part of the UL grant is used to transmit a Padding BSR MAC control element.

- Periodic BSR: This is triggered periodically according to the timer, periodicBSR-Timer, set by the RRC as in Table 6.1. Upon expiry of this timer, a periodic BSR MAC control element is transmitted.

The BSR timers are summarized in Table 6.1. A MAC PDU has at most one of the above three BSR MAC control element. A BSR control element always reflects the uplink buffer size at the UE. Different buffer sizes are indicated to eNodeB by their indices as in Table 6.7. The MAC control element for BSR is shown Figure 6.12.

Table 6.7   Buffer Sizes

Index	Buffer Size (Bytes)	Index	Buffer Size (Bytes)
0	BS = 0	32	$1132 < BS \leq 1326$
1	$0 < BS \leq 10$	33	$1326 < BS \leq 1552$
2	$10 < BS \leq 12$	34	$1552 < BS \leq 1817$
3	$12 < BS \leq 14$	35	$1817 < BS \leq 2127$
4	$14 < BS \leq 17$	36	$2127 < BS \leq 2490$
5	$17 < BS \leq 19$	37	$2490 < BS \leq 2915$
6	$19 < BS \leq 22$	38	$2915 < BS \leq 3413$
7	$22 < BS \leq 26$	39	$3413 < BS \leq 3995$
8	$26 < BS \leq 31$	40	$3995 < BS \leq 4677$
9	$31 < BS \leq 36$	41	$4677 < BS \leq 5476$
10	$36 < BS \leq 42$	42	$5476 < BS \leq 6411$
11	$42 < BS \leq 49$	43	$6411 < BS \leq 7505$
12	$49 < BS \leq 57$	44	$7505 < BS \leq 8787$
13	$57 < BS \leq 67$	45	$8787 < BS \leq 10287$
14	$67 < BS \leq 78$	46	$10287 < BS \leq 12043$
15	$78 < BS \leq 91$	47	$12043 < BS \leq 14099$
16	$91 < BS \leq 107$	48	$14099 < BS \leq 16507$
17	$107 < BS \leq 125$	49	$16507 < BS \leq 19325$
18	$125 < BS \leq 146$	50	$19325 < BS \leq 22624$
19	$146 < BS \leq 171$	51	$22624 < BS \leq 26487$
20	$171 < BS \leq 200$	52	$26487 < BS \leq 31009$
21	$200 < BS \leq 234$	53	$31009 < BS \leq 36304$
22	$234 < BS \leq 274$	54	$36304 < BS \leq 42502$
23	$274 < BS \leq 321$	55	$42502 < BS \leq 49759$
24	$321 < BS \leq 376$	56	$49759 < BS \leq 58255$
25	$376 < BS \leq 440$	57	$58255 < BS \leq 68201$
26	$440 < BS \leq 515$	58	$68201 < BS \leq 79846$
27	$515 < BS \leq 603$	59	$79846 < BS \leq 93479$
28	$603 < BS \leq 706$	60	$93479 < BS \leq 109439$
29	$706 < BS \leq 826$	61	$109439 < BS \leq 128125$
30	$826 < BS \leq 967$	62	$128125 < BS \leq 150000$
31	$967 < BS <= 1132$	63	$BS > 150000$

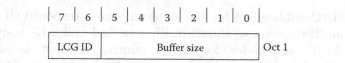

Figure 6.12: Buffer status report MAC control element.

6.6.3 Scheduling request

For the UE to get an UL grant, it should request one from the eNodeB. UE sends an SchedulingRequest (SR) for receiving an UL grant where the UE can transmit its MAC PDU. SR is not transmitted if logicalChannelSR-Prohibit timer is running. SR parameters are as shown in Table 6.1.

6.6.4 C-RNTI MAC control element

The UE uses this MAC control element to transmit its C-RNTI. It has a fixed size as shown in Figure 6.13. C-RNTI is a unique iden-tifier for the UE used by eNodeB for identifying RRC connection and DL/UL scheduling. After the RRC connection establishment or re-establishment procedures, the temporary C-RNTI is changed to become the C-RNTI after the contention resolution procedure passes successfully.

6.6.5 MAC PDU format

A MAC PDU is a single PDU that consists of a MAC header, a number of MAC SDUs, and a number of MAC control elements, and optionally a padding. The MAC header consists of a number of MAC subhead-ers. Each MAC subheader corresponds to a MAC SDU, MAC control element, or a padding.

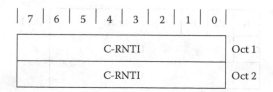

Figure 6.13: C-RNTI MAC control element.

The MAC subheaders for the random access response (RAR) and Back-off indicators are as shown in Figures 6.14 and 6.15, respectively.

The MAC subheader for MAC control element is shown in Figure 6.16 while Figure 6.17 shows the MAC subheader for a MAC SDU. The MAC subheader fields are summarized in Table 6.8.

The Logical Channel ID (LCID) is defined in Tables 6.9 and 6.10 for both downlink and uplink, respectively.

Figure 6.18 shows an example of one MAC PDU that contains two MAC control elements in addition to two MAC SDUs. Each MAC SDU contains an RLC PDU. For such a MAC PDU, there are five MAC subheaders (SH) which correspond to the two MAC Control Elements

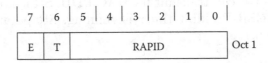

Figure 6.14: MAC subheader for RAR.

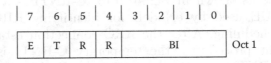

Figure 6.15: MAC subheader for back-off indicator.

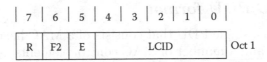

Figure 6.16: MAC subheader for MAC control element.

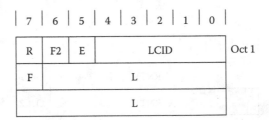

Figure 6.17: MAC subheader for MAC SDU.

Table 6.8   MAC Subheader Fields

Field	Possible Value	Meaning
LCID	Value from Table 6.9 or 6.10	Logical channel ID of MAC SDU or MAC control element
L	[1 32767]	Length of the MAC SDU or MAC control element
F	0, 1	Format field. Indicates the size of the Length field. If F is 1, size of MAC SDU or MAC control element is less than 128 otherwise it is zero
E	0, 1	Extension field. Indicates if more fields are present in the MAC subheader or not. If E field is set to 1, there is another set of at least R/F2/E/LCID fields otherwise, value of 0 indicates that either a MAC SDU, a MAC control element or padding starts at the next byte
T	0, 1	This only appplies to MAC PDU associated with RA-RNTI. A value of 1, indicates the MAC subheader has a RAPID (Random Access Preamble ID) otherwise, a value of 0 indicates a Back-off indicator in the MAC subheader
F2	0	Always set to zero
R	0	Reserved

Table 6.9   Downlink LCID

Index	LCID values
00000	CCCH
00001–01010	Identity of the logical channel
11100	UE Contention Resolution Identity
11101	Timing advance command
11110	DRX Command
11111	Padding

Table 6.10   Uplink LCID

Index	LCID values
00000	CCCH
00001–01010	Identity of the logical channel
11011	C-RNTI
11101	Short BSR
11111	Padding

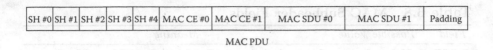

Figure 6.18: MAC PDU example.

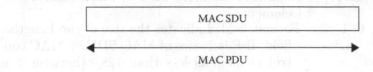

Figure 6.19: Transparent MAC PDU.

(CEs), two MAC SDUs, and Padding. Depending on the transport block size, there can be unused space left in the MAC PDU, in such, a padding is added at the end of the MAC PDU.

A transparent MAC PDU does not have any MAC subheader and contains only MAC SDU as shown in Figure 6.19. This MAC PDU is used by eNodeB for transmissions on PCH and BCH.

Chapter 7

Physical Sublayer

The physical sublayer is the bottom sublayer that is responsible for physical channels, transmissions, and receptions of MAC PDUs; it is illustrated in Figure 7.1. RRC provides the configuration parameters for the PHY subayer. At the MAC/PHY interface, transport channels are mapped to physical channels and vice-versa at transmission and reception, respectively [28].

RRC sends its configuration parameters to each sublayer, including the PHY sublayer, as indicated in Sections 4.2, 5.2, 6.2, and 7.1.

7.1 RRC Configuration Parameters

RRC sends dedicated or default radio configuration parameters to the PHY sublayer to be able to process transmissions and receptions on DL and UL for anchor carrier or non-anchor carrier. These configurations are shown in Table 7.1. The PHY configuration parameters are received by RRC from eNodeB during RRC connection establishment procedure as explained in Section 3.7.7.

7.2 FDD Frame Structure

The downlink frame structure is shown in Figure 7.2. Each radio frame is $T_f = 10$ ms long and consists of 20 slots of length $T_{slot} = 0.5$ ms, numbered from 0 to 19. A subframe is defined as two consecutive slots

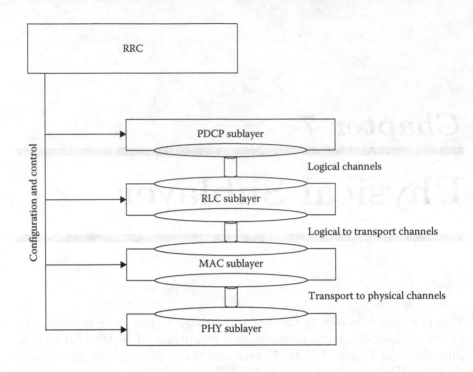

Figure 7.1: Description of PHY sublayer.

where subframe, n_{sf}, consists of slots $2n_{sf}$ and $2n_{sf} + 1$. Subframe, n_{sf}, in system frame number, n_f, has an absolute subframe number $n_{sf}^{abs} = 10n_f + n_{sf}$.

For uplink frame, the slot number within a radio frame is denoted n_s where $n_s \in \{0, 1, \ldots, 19\}$ for $\Delta f = 15$ KHz and $n_s \in \{0, 1, \ldots, 4\}$ for $\Delta f = 3.75$ KHz.

The uplink frame structure is the same as in Figure 7.2 with a duration of 10 ms and 20 slots if $\Delta f = 15$ KHz. If $\Delta f = 3.75$ KHz, the frame duration is 10 ms with only 5 slots (a slot is 2 ms) as shown in Figure 7.3 where the slot boundary is the same as subframe boundaries in Figure 7.2.

For half-duplex FDD, 20 slots are available for downlink transmission and either 20 slots ($\Delta f = 15$ KHz) or 5 slots ($\Delta f = 3.75$ KHz) are available for uplink transmissions in each 10 ms interval. Uplink and downlink transmissions are separated in the frequency domain. In half-duplex FDD operation, the UE cannot transmit and receive at the same time.

NB-IoT UE only supports half-duplex type B. In Type-B half-duplex FDD, a whole subframe is used as a guard between reception and

Table 7.1   RRC Configuration Parameters for PHY Sublayer

Parameter	Value	Meaning
carrierFreq	[0 262143]	Indicates the EARFCN for the NB-IoT carrier frequency. It can be used for DL or UL of a non-anchor carrier or UL of an anchor carrier.
carrierFreqOffset	v-10, v-9, v-8, v-7, v-6, v-5, v-4, v-3, v-2, v-1, v-0dot5, v0, v1, v2, v3, v4, v5, v6, v7, v8, v9	Offset of the NB-IoT channel number to EARFCN. It can be used for DL or UL of a non-anchor carrier or UL of an anchor carrier. Value v-10 means offset of −10
subframePattern10	10 bits	Downlink subframe configuration over 10 ms for inband, standalone, and guardband. Only used for non-anchor carrier The first/leftmost bit represents subframe #0 in radio frame where SFN mod 10 = 0. Value of 0 means that the subframe is invalid for DL transmission. Value 1 means the subframe is valid for DL transmission
subframePattern40	40 bits	Downlink subframe configuration over 40ms for inband. Only used for non-anchor carrier The first/leftmost bit represents subframe #0 in radio frame where SFN mod 40 = 0. Value of 0 means that the subframe is invalid for DL transmission. Value 1 means the subframe is valid for DL transmission

(Continued)

Table 7.1 (*Continued*)   RRC Configuration Parameters for PHY Sublayer

Parameter	Value	Meaning
indexToMidPRB	[−55 54]	Indicates an index to the Physical Resource Block (PRB) used for NB-IoT relative to the middle of the number of PRBs available. Only used for non-anchor carrier
ack-NACK-NumRepetitions	r1, r2, r4, r8, r16, r32, r64, r128	Indicates number of repetitions for the uplink ACK/NACK RU as a response to NPDSCH. r128 means 128 repetitions
ack-NACK-NumRepetitions-Msg4	r1, r2, r4, r8, r16, r32, r64, r128	Indicates number of repetitions for the uplink ACK/NACK RU for each NPRACH resource as a response to NPDSCH Msg4
twoHARQ-ProcessesConfig	True	Indicates whether a two HARQ processes are to be used or not
eutraControlRegion-Size	n1, n2, n3	Indicates the NPDCCH region size for inband operation mode in number of OFDM symbols
operationModeInfo	Inband-SamePCI, Inband-DifferentPCI, Guardband, Standalone	Inband-SamePCI: NB-IoT and eNodeB share the same physical cell ID Inband-DifferentPCI: NB-IoT and eNodeB have different physical cell ID Guardband: guardband deployment Standalone: standalone deployment

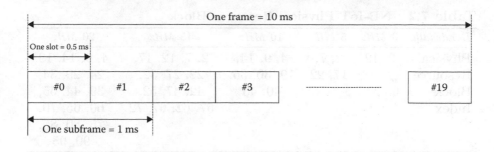

Figure 7.2: Downlink frame structure and uplink frame structure for $\Delta f = 15$ KHz.

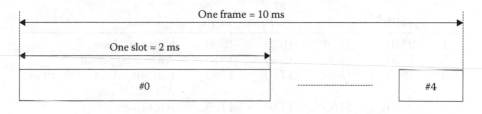

Figure 7.3: Uplink frame structure for $\Delta f = 3.75$ KHz.

transmission. This allows for the low-cost implementation with only a single oscillator that can be switched between uplink and downlink frequencies.

A UE operates in the downlink using 12 subcarriers with a subcarrier bandwidth of 15 KHz (a bandwidth of 180 KHz), and in the uplink using a single subcarrier with a subcarrier bandwidth of either 3.75 or 15 KHz or alternatively 3, 6, or 12 subcarriers with a subcarrier bandwidth of 15 KHz.

In the frequency domain in each slot, there is only one resource block per each NB-IoT carrier. There can be more than one NB-IoT carrier configured as in Table 3.31.

Table 7.2 shows a proposed resource block allocation to be used by NB-IoT for inband mode where eNodeB uses only one resource block per carrier [29].

7.3 Channel Frequency Band

NB-IoT devices use specific bands and frequencies for their DL and UL transmissions. The frequency bands used by NB-IoT devices are as shown in Table 7.3 [30, 31].

Table 7.2   NB-IoT Physical Resource Block

Bandwidth	3 MHz	5 MHz	10 MHz	15 MHz	20 MHz
Physical Resource Block Index	2, 12	2, 7, 17, 22	4, 9, 14, 19, 30, 35, 40, 45	2, 7, 12, 17, 22, 27, 32, 42, 47, 52, 57, 62, 67, 72	4, 9, 14, 19, 24, 29, 34, 39, 44, 55, 60, 65, 70, 75, 80, 85, 90, 95

Table 7.3   Channel Frequency Band

Band	F_{DL}^{low} (MHz)	F_{DL}^{high} (MHz)	F_{UL}^{low} (MHz)	F_{UL}^{high} (MHz)	Region
	Downlink		Uplink		
1	2110	2170	1920	1980	Europe, Asia
2	1930	1990	1850	1910	Americas, Asia
3	1805	1880	1710	1785	Europe, Asia, Americas, Africa
4	2110	2155	1710	1755	Americas
5	869	894	824	849	Americas, Asia
8	925	960	880	915	Europe, Asia, Africa
11	1844.9	1879.9	1749.9	1784.9	Japan
12	729	746	699	716	United States
13	746	756	777	787	United States
14	758	768	788	798	United States
17	734	746	704	716	United States
18	860	875	815	830	Japan
19	875	890	830	845	Japan
20	791	821	832	862	Europe, Africa
21	1495.9	1510.9	1447.9	1462.9	Europe
25	1930	1995	1850	1915	Americas
26	859	894	814	849	Americas, Japan
28	758	803	703	748	Americas, Asia Pacific
31	462.5	467.5	452.5	457.5	Americas
66	2110	2200	1710	1780	Americas
70	1995	2020	1695	1710	Americas
71	461	466	451	456	United States
72	460	465	450	455	Europe, Middle East, Africa
73	1475	1518	1427	1470	Asia and Pacific

The frequency bands defined for NB-IoT are paired bands for FDD duplex mode only. NB-IoT supports FDD duplex mode only and not TDD.

7.4 Carrier Frequency

E-UTRA Absolute Radio Frequency Channel Number (EARFCN) designates the carrier frequency used by NB-IoT device. It is given by the equation:

$$F_{DL} = F_{DL}^{low} + 0.1 \times (N_{DL} - N_{DL}^{Off}) + 0.0025 \times (2 \times M_{DL} + 1),$$

where F_{DL} is the downlink carrier frequency. F_{DL}^{low} is given in Table 7.3. N_{DL} is the downlink EARFCN and N_{DL}^{Off} is an offset. Both N_{DL} and N_{DL}^{Off} are as shown in Table 7.4. M_{DL} is the Offset channel number to downlink EARFCN in the range $\{-10,-9,-8,-7,-6,-5,-4,-3,-2,-1, -0.5,0,1,2,3,4,5,6,7,8,9\}$.

EARFCN for the uplink is given by:

$$F_{UL} = F_{UL}^{low} + 0.1 \times (N_{UL} - N_{UL}^{Off}) + 0.0025 \times (2 \times M_{UL}),$$

Table 7.4 EARFCN

Band	Downlink N_{DL}^{Off}	Range of N_{DL}	Uplink N_{UL}^{Off}	Range of N_{UL}
1	0	0–599	18000	18000–18599
2	600	600–1199	18600	18600–19199
3	1200	1200–1949	19200	19200–19949
4	1950	1950–2399	19950	19950–20399
5	2400	2400–2649	20400	20400–20649
8	3450	3450–3799	21450	21450–21799
11	4750	4750–4949	22750	22750–22949
12	5010	5010–5179	23010	23010–23179
13	5180	5180–5279	23180	23180–23279
14	5280	5280–5379	23280	23280–23379
17	5730	5730–5849	23730	23730–23849
18	5850	5850–5999	23850	23850–23999
19	6000	6000–6149	24000	24000–24149
20	6150	6150–6449	24150	24150–24449
21	6450	6450–6599	24450	24450–24599
25	8040	8040–8689	26040	26040–26689
26	8690	8690–9039	26690	26690–27039
28	9210	9210–9659	27210	27210–27659
31	9870	9870–9919	27760	27760–27809
66	66436	66436–67335	131972	131972–132671
70	68336	68336–68585	132972	132972–133121
71	68586	68586–68935	133122	133122–133471
72	68936	68936–68985	133472	133472–133521
73	68986	68986–69035	133522	133522–133571

where F_{UL} is the uplink carrier frequency. F_{UL}^{low} is given in Table 7.3. N_{UL} is the uplink EARFCN and N_{UL}^{Off} is an offset. Both N_{UL} and N_{UL}^{Off} are as shown in Table 7.4. M_{UL} is the Offset Channel Number to uplink EARFCN in the range $\{-10,-9,-8,-7,-6,-5,-4,-3,-2,-1, 0,1,2,3,4,5,6,7,8,9\}$.

N_{DL}, N_{UL}, M_{DL}, and M_{UL} are advertised to the UE (carrierFreq and carrierFreqOffset parameters) as shown in Table 7.1.

For standalone operation, only $M_{DL} = -0.5$ is applicable. $M_{DL} = -0.5$ is not applicable for inband and guardband operations. For the carrier including NPSS/NSSS for inband and guardband operation, M_{DL} is selected from $\{-2,-1,0,1\}$.

7.5 Downlink and Uplink Channel Frequency Separation

There is a separation between the downlink (Rx) and uplink frequency (Tx) as shown in Table 7.5.

7.6 Carrier Frequency Raster

Carrier raster refers to the frequency separation between carrier center frequencies. For each supported band, an NB-IoT carrier can exist on each 100 KHz carrier raster or carrier separation. This can be expressed as $n \times 100$ KHz where n is an integer. If the NB-IoT UE is in IDLE mode and is turned on, it can search for DL frequencies assuming a raster of 100 KHz.

7.7 Channel and Transmission Bandwidth

NB-IoT UE uses a specific channel or transmission bandwidth depending on the mode of operation (standalone, inband, or guardband) [31]. For NB-IoT standalone operation, both channel and transmission bandwidths are 200 KHz and 180 KHz, respectively. This is illustrated in Figure 7.4 and Table 7.6 where only one resource block is available for NB-IoT transmission [32].

The frequency domain waveform of the 200 KHz resource block channel bandwidth is shown in Figure 7.5.

For inband operation, channel bandwidth can range from 3 to 20 MHz as shown in Table 7.7. Transmission bandwidth is 180 KHz.

Table 7.5 Tx and Rx Frequency Separation

Band	Separation between DL and UL Frequency
1	190
2	80
3	95
4	400
5	45
8	45
11	48
12	30
13	−31
14	−30
17	30
18	45
19	45
20	−41
21	48
25	80
26	45
28	55
31	10
66	400
70	300
71	−46
72	10
73	10

Relation between channel and transmission bandwidth is as illustrated in Figure 7.6 where only one resource block is used for transmission.

For guardband operation, it uses the same channel bandwidth as in Figure 7.7 except that channel bandwidth of 3 MHz is not used for this mode of operation. Transmission bandwidth is 180 KHz. Relation between channel and transmission bandwidth is as illustrated in Figure 7.7 where only one resource block is used for transmission.

7.8 Mapping of Physical Channels

Transport channels, at the MAC sublayer, are mapped to physical channels at the PHY sublayer. Figure 7.8 shows the mapping of transport channels to/from physical channels.

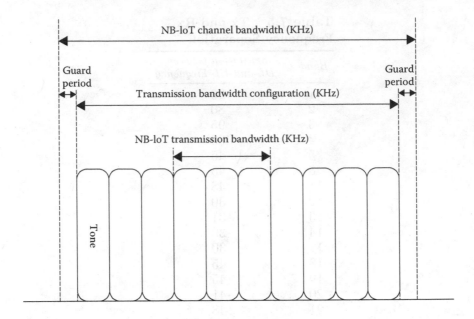

Figure 7.4: Channel bandwidth and transmission bandwidth for standalone operation.

Table 7.6   Channel Bandwidth for Standalone Operation

Characterstic	Value
Channel bandwidth (KHz)	200
Number of downlink resource block (N_{RB})	1
Number of uplink subcarriers for $\Delta f = 15$ KHz	12
Number of uplink subcarriers for $\Delta f = 3.75$ KHz	48

7.9   Physical Cell ID (PHY_{CELL}^{ID})

Physical cell ID is the first parameter the UE must obtain during a cell search in order to be able to decode or encode downlink and uplink physical channels.

If RRC signals operationModeInfo as "inband-SamePCI" for a cell, the UE assumes that the physical layer cell ID is same as the narrowband physical layer cell ID for the cell.

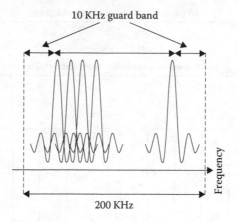

Figure 7.5: Frequency-domain of 200 KHz channel bandwidth.

Table 7.7   Channel Bandwidth for Inband and Guardband Operations

Channel bandwidth (MHz)	3	5	10	15	20
Transmission bandwidth (N_{RB})	15	25	50	75	100

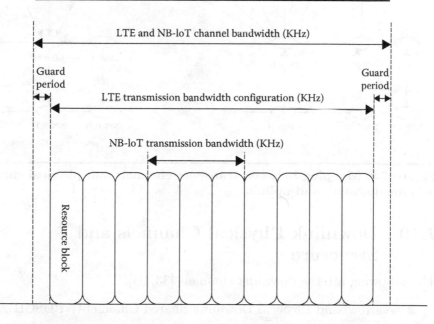

Figure 7.6: Channel bandwidth and transmission bandwidth for inband operation.

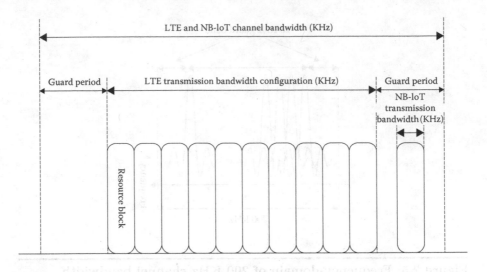

Figure 7.7: Channel bandwidth and transmission bandwidth for guardband operation.

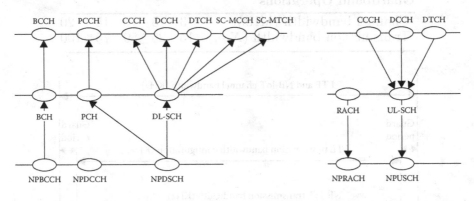

Figure 7.8: Mapping between transport channels and physical channels for downlink and uplink.

7.10 Downlink Physical Channels and Structure

The following are the downlink channels [33, 34]:

■ Narrowband Physical Downlink Shared Channel (NPDSCH).

■ Narrowband Physical Broadcast Channel (NPBCH).

■ Narrowband Physical Downlink Control Channel (NPDCCH).

In addition, the following signals are defined:

- Narrowband Reference Signal (NRS).

- Narrowband Synchronization Signal.

- Narrowband Positioning Reference Signal (NPRS).

7.10.1 Downlink transmission scheme OFDM

The downlink transmission scheme is based on conventional OFDM using a cyclic prefix. The OFDM subcarrier spacing of $\Delta f = 15$ KHz is only supported. In the frequency domain, 12 consecutive subcarriers during one slot correspond to one downlink resource block. In the time domain, the number of resource blocks is one and only one resource block per slot that is assigned for an NB-IoT UE. In the case of 15 KHz subcarrier spacing, there are two cyclic-prefix lengths, corresponding to seven and six OFDM symbols per slot, respectively.

- Normal cyclic prefix: $T_{CP} = 5.2$ us (OFDM symbol #0), $T_{CP} = 4.7$ us (OFDM symbol #1 to #6).

- Extended cyclic prefix: $T_{CPe} = 16.67$ us (OFDM symbol #0 to OFDM symbol #5).

Figure 7.9 shows the OFDM symbol which consists of the cyclic-prefix, T_{CP}, and the useful symbol duration, T_u, where $T_u = 1/\Delta f$.

7.10.2 Resource grid

Downlink resource grid consists of 7 or 6 OFDM symbols in time domain and 12 subcarriers in frequency domain. This is shown in Figure 7.10 and its parameters are summarized in Table 7.8. Each resource element represents a single subcarrier.

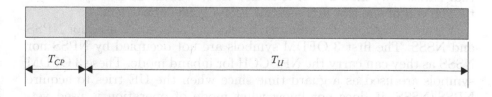

Figure 7.9: OFDM symbol.

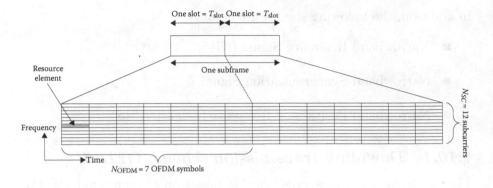

Figure 7.10: Downlink resource grid.

Table 7.8   Downlink Slot

Cyclic Prefix	N_{OFDM}	Subcarrier Spacing	N_{sc}
Normal	7	$\Delta f = 15$ KHz	12
Extended	6	$\Delta f = 15$ KHz	

7.10.3   Primary and secondary synchronization signals

Narrowband Primary Synchronization Signal (NPSS) and Narrowband Secondary Synchronization Signal (NSSS) are the first signals to be acquired by the UE. When the UE is powered on or an USIM is inserted, the UE starts to scan all the RF channels available to find a suitable cell. The UE acquires the NPSS and NSSS in order to obtain the physical cell ID, PHY_{CELL}^{ID}, and be able to decode and encode downlink and uplink physical channels. There are possible 504 physical cell IDs and UE uses the NPSS/NSSS to determine the physical cell ID. eNodeB transmits NPSS in subframe #5 in each radio frame and is located in OFDM symbol #3 till the end of the subframe and starts from subcarrier #0 to subcarrier #10. NSSS is transmitted in subframe #9 in radio frames that satisfies $n_f \bmod 2 = 0$. It is also located from OFDM symbol #3 till the end of the slot and uses all 12 assigned subcarriers.

Figure 7.11 shows the resource blocks and subcarrier used for NPSS and NSSS. The first 3 OFDM symbols are not occupied by NPSS nor NSSS as they can carry the NPDCCH for inband mode. These 3 OFDM symbols are used as a guard time since when the UE tries to acquire NPSS/NSSS, it does not know what mode of operation is used yet. NPSS and NSSS, are only transmitted in a resource block as indicated in Table 7.2.

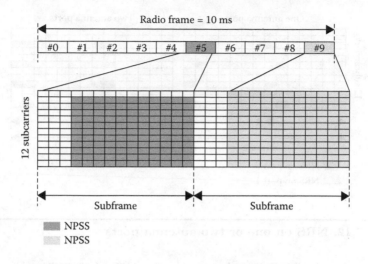

Figure 7.11: NSSS and NPSS resource blocks.

7.10.4 Reference and cell-specific reference signals

NRS is received by UE and can be used for downlink power control or channel estimation.

If the UE has not acquired the operationModeInfo,

■ NRS is transmitted in subframes #0 and #4 and in subframes #9 not containing NSSS.

If UE acquired operationModeInfo indicating guardband or standalone,

■ NRS is transmitted in subframes #0, #1, #3, #4, subframes #9 not containing NSSS, and in all DL subframes assigned for NB-IoT.

If UE acquired operationModeInfo indicating inband,

■ NRS is transmitted in subframes #0, #4, subframes #9 not containing NSSS, and in all DL subframes assigned for NB-IoT.

NRS is not transmitted in subframes containing NPSS or NSSS. Figure 7.12 shows the NRS transmitted on assigned subcarriers and subframes when one or two antenna ports are used.

Cell-Specific Reference (CSR) signal is transmitted in DL subframes where the NRS is available and using the same number of antenna ports (either one or two antenna ports) used by NRS. Subcarrier assigned

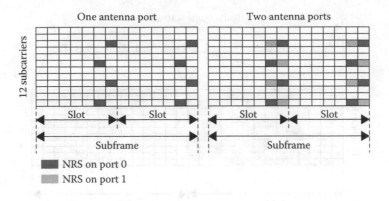

Figure 7.12: NRS on one or two antenna ports.

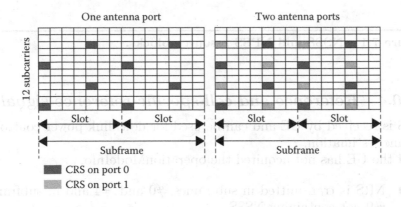

Figure 7.13: CRS on one or two antenna ports.

to CSR are additionally cyclically shifted by PHY_{CELL}^{ID} mod 6 in the frequency range. If the subframes and assigned subcarriers are used for NPSS/NSSS, CRS punctures the NPSS/NSSS in those subframes. CRS is shown in Figure 7.13 for both single and two antenna ports.

7.10.5 Downlink power control

The eNodeB controls and calculates the power used for downlink subcarriers and subframes. DL transmission power refers to the NRS transmission power. The value of the NRS transmission power is transmitted to the UE and UE uses it to calculate and estimate the path loss.

NRS transmission power, NRS Energy Per Resource Element (EPRE), is constant across all DL subcarriers and subframes. NRS

EPRE is calculated as the linear average over the power contributions (in [W]) of all subcarriers that carry NRS. For the NPBCH, NPDCCH, and NPDSCH, the transmit power depends on the transmission scheme. If only one antenna port is applied, the power is the same as for the NRS; otherwise, it is reduced by 3 dB.

A special case applies if the inband operation mode is used and the samePCI value is set to true. In this case, the eNodeB can additionally signal the ratio of the NRS power to the CRS power which enables the UE to use the CRS for channel estimation.

7.10.6 *Modulation schemes*

Each of the physical downlink channel uses a modulation scheme. Those schemes are summarized in Table 7.9.

7.10.7 *NPBCH*

This is the physical channel used for receiving broadcast control PDU, MIB-NB, from the eNodeB. The size of the MIB-NB is 34 bits and its transport block is transmitted each Transmission Time Interval (TTI) of 640 ms. The Cyclic Redundancy Check (CRC) of PBCH is scrambled with a 16-bit CRC mask according to whether 1 or 2 antenna ports are used as in Table 7.10.

After adding CRC to the transport block, channel coding, and rate matching, it results in a number of 1600 bits. Since QPSK has 2 bits constellation size, it results in 800 symbols to be transmitted to the UE. NPBCH is transmitted in subframe #0 during 64 consecutive radio

Table 7.9 Modulation Scheme Used for Physical Downlink Channels

Physical Channel	Modulation Scheme
NPBCII	QPSK
NPDSCH	QPSK
NPDCCH	QPSK

Table 7.10 NPBCH CRC Mask

Number of Antenna Port	CRC Mask
1	0, 0, 0, 0, 0, 0, 0, 0, 0, 0, 0, 0, 0, 0, 0, 0
2	1, 1, 1, 1, 1, 1, 1, 1, 1, 1, 1, 1, 1, 1, 1, 1

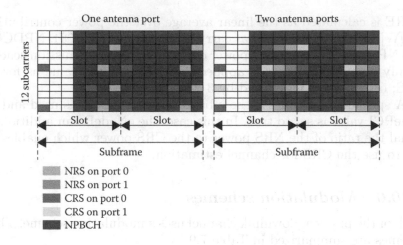

One antenna port Two antenna ports

12 subcarriers

Slot   Slot   Slot   Slot

Subframe   Subframe

- NRS on port 0
- NRS on port 1
- CRS on port 0
- CRS on port 1
- NPBCH

Figure 7.14: NPBCH on subframe #0 and with NRS and CRS for single or two antenna ports.

frames starting in a radio frame that fulfils n_f mod 64 = 0. Modulation is used as in Table 7.9.

The 800 symbols are divided into 8 blocks where each block has 100 symbols. The 64 consecutive radio frames are grouped into 8 radio group where each group has 8 radio frames. In each radio group, subframe #0 in the first radio frame is used to transmit one block and the subsequent subframes, subframe #0, in the same radio group contains the repetitions of the same block.

Figure 7.14 shows the allocated subcarriers assigned to NPBCH. In subframe #0, there are NRS and CSR. In addition, the first 3 OFDM symbols are not occupied by NPBCH. NPBCH can be transmitted using a single or two antenna ports.

7.10.8 NPDSCH

NPDSCH is used to carry traffic for DL-SCH and PCH. NPDSCH carries only one resource block for a UE per subframe. NPDSCH carries a transport block which contains one full MAC PDU. Only type-B half-duplex FDD operation is supported. Only a subframe designated as an NB-IoT downlink subframe contains the resource block allocated for the UE. UE assumes a subframe is NB-IoT subframe if:

- The UE determines that the subframe does not contain NPSS/NSSS/NPBCH/NB-SIB1 transmission, and

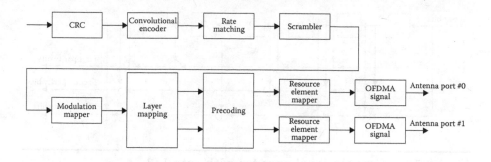

Figure 7.15: NPDSCH physical layer processing.

■ The subframe is configured as NB-IoT DL subframe after the UE has obtained SystemInformationBlockType1-NB.

The NPDSCH performs the following as in Figure 7.15:

■ CRC attachment.

■ Channel coding: Convolutional coding.

■ Rate matching.

■ Scrambling.

■ Modulation.

■ Layer mapping and pre-coding.

■ Mapping to assigned resources and antenna ports.

7.10.8.1 CRC Calculation

CRC provides error detection capability for transport block transmitted on the downlink. If the transport block has S bits, additional P bits corresponds to the CRC are concatenated to the transport block bits. The parity bits are computed and attached to the transport block using the generator polynomial $G_{24}(X)$ where

$$G_{24}(X) = X^{24} + X^{23} + X^{18} + X^{17} + X^{14} + X^{11} + X^{10} + X^7 + X^6 + X^5$$
$$+ X^4 + X^3 + X + 1.$$

The bits including transport block bits and CRC bits are denoted by $b_0, b_1, b_2, \ldots, b_{B-1}$, where $B = S + P$.

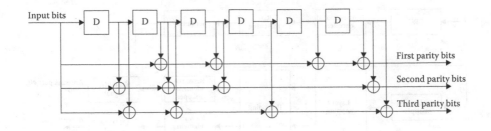

Figure 7.16: Convolutional encoder with rate 1/3.

7.10.8.2 Convolutional Coding

The output from the CRC attachment is one code block denoted by $c_0, c_1, c_2, \ldots, c_{K-1}$, where K is the number of bits for the code block. There is only one code block since the number of inputs bits is less than the code block, $Z = 6144$. The code block is convolutionally encoded. A tail biting convolutional encoder with constraint length 7 and coding rate 1/3 is used as shown in Figure 7.16.

NPDSCH uses tail-biting convolutional coding and not Turbo coding (as in legacy LTE™) since this results in a lower channel coding complexity at the UE side.

The output from the convolutional encoder are the parity bits denoted by $p_0^i, p_1^i, p_2^i, \ldots, p_{K-1}^i$, where $i = 0, 1, 2$ and $K = B$.

7.10.8.3 Rate Matching

The output from the convolutional encoder is provided as an input to the rate matching block shown in Figure 7.17. The three information bit streams, p_k^0, p_k^1, p_k^2, obtained from the convolutional encoder, are provided as an input to each subblock interleaver which interleaves them separately. The bit selection selects output bits of a length equal to E. The output sequence from the rate matching is denoted by $e_0, e_1, e_2, \ldots, e_{E-1}$, where E is the number of rate matched bits. To keep the NB-IoT UE complexity low, only a single redundancy version (RV) is specified for NPDSCH.

This sequence of coded bits, which correspond to one transport block after the rate matching, is referred to as a *codeword*.

7.10.8.4 Scrambling

The input to the bit-level scrambler are $e_0, e_1, e_2, \ldots, e_{E-1}$, where E is the number of bits to be transmitted, is scrambled before modulated.

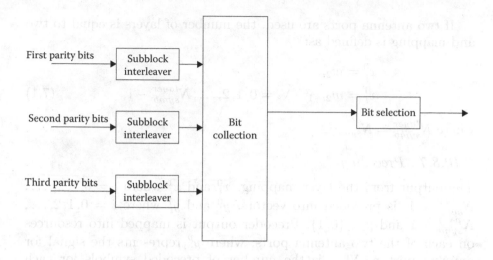

Figure 7.17: Rate matching for convolutional encoder with rate 1/3.

The scrambler output is $h_0, h_1, h_2, \ldots, h_{E-1}$ where

$$h_i = (e_i + scr_i) \bmod 2.$$

scr_i is a constant scrambling sequence which depends on C-RNTI and physical cell ID (it is a UE-specific scrambling sequence).

7.10.8.5   Modulation

The output from the scrambler, $h_0, h_1, h_2, \ldots, h_{E-1}$, is modulated using QPSK resulting in a block of complex modulation symbols, $m_0, m_1, m_2, \ldots, m_{N_{sym}-1}$, where N_{sym} is the number of complex modulation symbols. Modulation scheme used for NPDSCH is summarized in Table 7.9.

7.10.8.6   Layer Mapping

Modulation symbols are mapped into one or two layers. Complex-valued modulation symbols $m_0, m_1, m_2, \ldots, m_{N_{sym}-1}$ are mapped into a maximum of two layers, x_i^0 and x_i^1, where $i = 0, 1, 2, \ldots, N_{symb}^{layer} - 1$ and N_{symb}^{layer} is the number of modulation symbols per layer.

If a single antenna port is used on the downlink, a single layer is used, and the mapping is defined as

$$x_i^0 = m_i, \quad \forall i = 0, 1, 2, \ldots, N_{symb}^{layer} - 1,$$

where $N_{symb}^{layer} = N_{symb}$.

If two antenna ports are used, the number of layers is equal to two and mapping is defined as:

$$x_i^0 = m_{2i},$$
$$x_i^1 = m_{2i+1}, \quad \forall i = 0, 1, 2, \ldots, N_{symb}^{layer} - 1, \qquad (7.1)$$

where $N_{symb}^{layer} = N_{symb}/2$.

7.10.8.7 Precoding

The output from the layer mapping, x_i^0 and x_i^1 where $i = 0, 1, 2, \ldots,$ $N_{symb}^{layer} - 1$, is precoded into vectors, y_i^0 and y_i^1 where $i = 0, 1, 2, \ldots,$ $N_{symb}^p - 1$ and $p \in \{0, 1\}$. Precoder output is mapped into resources on each of the two antenna ports, where y_i^p represents the signal for antenna port p. N_{sym}^p is the number of precoded symbols for each antenna port. If a single antenna port is used, $p = 0$, precoding is defined by:

$$y_i^0 = x_i^0, \quad \forall i = 0, 1, 2, \ldots, N_{symb}^{layer} - 1.$$

If two antenna ports are used (for transmit diversity), the output is based on a space-frequency block coding (SFBC). SFBC implies that two consecutive modulation symbols y_{2i}^0 and y_{2i+1}^0 are mapped directly to frequency-adjacent resource elements on the first antenna port. On the second antenna port, the symbols $-y_{2i+1}^{*0}$ and y_{2i}^{*0} are mapped to the corresponding resource elements[1]. For two antenna ports, $p \in \{0, 1\}$, the output y_i^0 and y_i^1 of the precoding operation can be defined as

$$\begin{bmatrix} y_{2i}^0 \\ y_{2i}^1 \\ y_{2i+1}^0 \\ y_{2i+1}^1 \end{bmatrix} = \frac{1}{\sqrt{2}} \begin{bmatrix} 1 & 0 & j & 0 \\ 0 & -1 & 0 & j \\ 0 & 1 & 0 & j \\ 1 & 0 & -j & 0 \end{bmatrix} \begin{bmatrix} Re(x_i^0) \\ Re(x_i^1) \\ Im(x_i^0) \\ Im(x_i^1) \end{bmatrix}, \quad \forall i = 0, 1, 2, \ldots, N_{symb}^{layer} - 1,$$

where $N_{symb}^p = 2N_{symb}^{layer}$.

Figure 7.18 shows the mapping of the modulations symbols on the two layers to the two antenna ports for transmit diversity.

7.10.8.8 Mapping to Physical Resources

The output from the precoder, y_i^0 and y_i^1, are mapped to a subcarrier of a resource block starting with the first slot and then second slot

---

[1] "*" denotes complex conjugate.

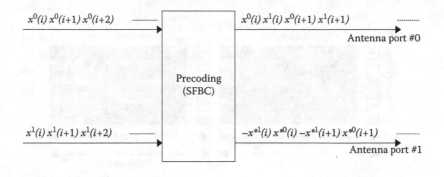

Figure 7.18: Transmit diversity.

of a downlink subframe. Each subframe contains a maximum of 168 subcarriers (12 subcarrier in 14 OFDM symbols). Some subcarriers are used for NPSS/NSSS, NPDBCH, NPDCCH, or NRS and cannot be used for NPDSCH.

Repetition of the subframe is a technique used to repeat the same subframe several times. Repetition increases the coverage (up to 20 dB) and all repetitions are ACKed just once. For each subframe, it is repeated a number of times before continuing the mapping of y_i^0 and y_i^1 to other subcarriers in another subframe.

7.10.8.9 NPDSCH Position and Mapping

NPDSCH can carry broadcast system information (SIBs) or a unicast traffic. For unicast traffic, NPDSCH starts in OFDM symbol, $l_{DataStart}$, located in the first slot in a subframe where $l_{DataStart}$ is equal to eutraControlRegionSize, as in Table 7.1, if provided by RRC otherwise, $l_{DataStart} = 0$. If NPDSCH carries SIB1-NB, $l_{DataStart} = 3$ for inband operation otherwise $l_{DataStart} = 0$ for guardband or standalone operations.

The start OFDM symbol avoids conflict with the LTE PDCCH channel if NB-IoT is inband operation. However, for both guardband and standalone operation modes, the start of the NPDSCH always starts from OFDM symbol #0 which provides more resource blocks space for NPDSCH.

Figure 7.19 shows the assigned subcarriers and OFDM symbols used for NPDSCH assuming inband operation and $l_{DataStart}$ is set equal to OFDM symbol #2. NPDSCH is not transmitted in subframe #0 as this is used for NPBCH. NPDSCH is only transmitted in DL subframes designated for NB-IoT transmission.

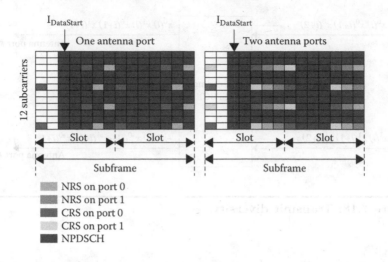

Figure 7.19: NPDSCH with NRS and CRS for single or two antenna ports for inband operation.

7.10.9 NPDCCH

The NPDCCH carries control information and does not carry any control-plane or data-plane PDU. DL and UL assignments and grants are indicated on NPDCCH.

NPDCCH supports aggregations of 1 or 2 consecutive Narrowband Control Channel Elements (NCCE) in a subframe and repetition. NCCE occupies 6 consecutive subcarriers in a subframe where NCCE 0 occupies subcarriers 0 through 5 and NCCE 1 occupies subcarriers 6 through 11. There are two NPDCCH formats as summarized in Table 7.11.

NPDCCH supports C-RNTI, Temporary C-RNTI, P-RNTI, RA-RNTI, SC-RNTI, and G-RNTI. The RNTI is implicitly encoded in the CRC by scrambling the calculated CRC by the RNTI. Figure 7.20 illustrates how the NPDCCH is scrambled with the RNTI.

One or two NPDCCHs can be transmitted in a subframe. In case of multiple carriers, multiple NPDCCHs from each carrier are multiplexed

Table 7.11   NPDCCH Format

NPDCCH Format	Number of NCCEs
0	1
1	2

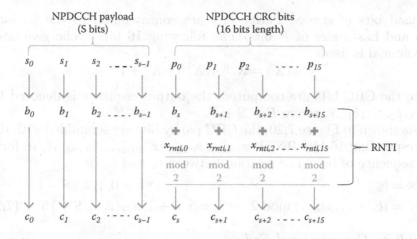

Figure 7.20: NPDCCH encoding by scrambling with RNTI.

together. The UE monitors multiple NPDCCH. Each NPDCCH has its own set of x-RNTI.

When the UE detects NPDCCH with Downlink Control Information (DCI) format N1, N2 ending in subframe n intended for the UE, it decodes, starting in $n + 5$ DL subframe, the corresponding NPDSCH transmission in a number of consecutive NB-IoT DL subframe(s).

The NPDCCH performs the following and the same blocks as shown in Figure 7.15.

- CRC insertion: 16-bit CRC.

- Channel coding: Tail biting convolutional coding.

- Rate matching.

- Channel Interleaving.

- Scrambling.

- Modulation.

- Layer mapping and pre-coding.

- Mapping to assigned resources and antenna ports.

7.10.9.1 CRC Calculation

CRC provides error detection capability for NPDCCH information and is also scrambled by RNTI. The input to this block is a number of payload bits of size S. Parity bits are concatenated to the payload bits and has a size of P bits. The following 16 bits cyclic generator polynomial is used:

$$G_{16}(X) = X^{16} + X^{12} + X^5 + 1.$$

Once the CRC bits are computed, the output sequence is denoted by $b_0, b_1, b_2, \ldots, b_{B-1}$, where $B = S + P$.

As shown in Figure 7.20, the CRC parity bits are scrambled with the corresponding 16-bits RNTI, $x_{rnti,0}, x_{rnti,1}, x_{rnti,2}, \ldots, x_{rnti,15}$, to form the sequence of bits. The relation between c_k and b_k is:

$$c_k = b_k \qquad\qquad\qquad\qquad \forall k = 0, 1, 2, , S - 1$$

$$c_k = (b_k + x_{rnti,k-s}) \bmod 2 \quad \forall k = S, S+1, S+2, ..., S+15. \quad (7.2)$$

7.10.9.2   Convolutional Coding

The output from the CRC block is convolutionally encoded the same way as in Section 7.10.8.2.

7.10.9.3   Rate Matching

This is the same block as used in Section 7.10.8.3 for NPDSCH. There is only one code block and there is no RV functionality.

7.10.9.4   Scrambling

This is the same block as used in Section 7.10.8.4 for NPDSCH. The scrambled bits are provided as an input to the modulator.

7.10.9.5   Modulation

The output from the scrambler is modulated resulting in a block of complex-valued modulation symbols, $m_0, m_1, m_2, \ldots, m_{N_{sym}-1}$, where N_{sym} is the number of modulation symbols. Modulation is done according to Table 7.9.

7.10.9.6   Layer Mapping

The same as used in Section 7.10.8.6 for NPDSCH.

7.10.9.7   Precoding

The same as used in Section 7.10.8.7 for NPDSCH.

7.10.9.8   Mapping to Resource Element

The complex-valued symbols, $m_0, m_1, m_2, \ldots, m_{N_{sym}-1}$ are mapped to a subcarrier in NCCE only in the downlink subframe designated for

NB-IoT. Modulation symbols are mapped to the OFDM symbols according to the NPDCCH position as indicated by the parameter $l_{NPDCCHStart}$.

7.10.9.9 NPDCCH Position and Mapping

NPDCCH is always located in the first slot in a DL subframe. The starting OFDM symbol where NPDCCH is located, $l_{NPDCCHStart}$, is given by:

- $l_{NPDCCHStart}$ is equal to eutraControlRegionSize if operationModeInfo indicates inband operation.

- $l_{NPDCCHStart}$ is equal to zero if operationModeInfo indicates guardband or standalone operations.

The start OFDM symbol avoids conflict with the LTE PDCCH channel if NB-IoT is inband operation. However, for both guardband and standalone operation modes, the start of the NPDCCH always starts from 0 which provides more resource blocks space for NPDCCH.

Figure 7.21 shows the assigned subcarriers and OFDM symbols used for NPDCCH assuming inband operation and $l_{NPDCCHStart}$ is set equal to OFDM symbol #2. NPDCCH is not transmitted in subframe #0 as this is used for NPBCH. NPDCCH is only transmitted in DL subframes designated for NB-IoT transmission.

NPDCCH assigned resources shown in Figure 7.21 is used by both NCCEs (NCCE0 and NCCE1). If NPDCCH format 0 is used, it takes one NCCE. If NPDCCH format 1 is used, it takes both NCCEs.

7.10.9.10 RRC Information for DCI Processing

Table 7.1 shows the RRC layer configuration parameters provided to the PHY layer to aid in the encoding and decoding of the DCI. Both parameters, operationModeInfo and eutraControlRegionSize, are necessary information for decoding the NPDCCH.

7.10.9.11 DCI Format N0

This format is used to indicate an uplink grant in the NPUSCH for a single uplink carrier. The information carried in DCI format N0 is shown in Table 7.12. The total size of DCI format N0 is 24 bits [35].

When the UE detects format N0 that ends in DL subframe, n, the NPUSCH, using format 1, starts in UL slots after the $(n + k)$th DL subframe where $8 \leq k \leq 64$. The NPUSCH is also repeated in the N

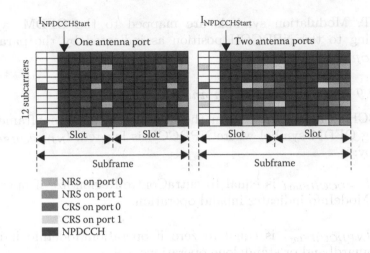

Figure 7.21: NPDCCH with NRS and CRS for single or two antenna ports for inband operation.

consecutive UL slots starting after the $(n+k)$th DL subframe. In other words:

- NPUSCH uses a subcarrier spacing of $\Delta f = 3.75$ KHz or $\Delta f = 15$ KHz which is determined by the UE when it receives Random Access Response (RAR) grant message as in Tables 6.4 and 6.5.

- NPUSCH is repeated in a number of consecutive UL slots where the number of repetitions is $N = N_{Rep} \times N_{RU} \times N_{slots}^{UL}$. N_{Rep} and N_{RU} are determined from the repetition number, I_{Rep}, and resource assignment, I_{RU}, fields, respectively, according to Tables 7.12, 7.14, and 7.13. N_{slots}^{UL} is the number of UL slots of the RU as in Table 7.29.

The NPDCCH transmission ends in subframe, n. NPUSCH transmission can be delayed a number of subframes k. That is, after subframe n ends, NPUSCH subframes starts k subframe after the subframe n and repeated N consecutive subframes. k is determined by the scheduling delay field, I_{Delay}, as in Tables 7.12 and 7.15.

To determine the UL modulation order and transport block size, the UE does the following:

- Each RUresource unit (RU) has a number of contiguously allocated subcarriers, n_{sc}, determined by subcarrier indication, I_{sc}, as in Table 7.12. For $\Delta f = 15$ KHz, n_{sc} is determined according to Table 7.16. For subcarrier spacing, $\Delta f = 3.75$ KHz, $n_{sc} = I_{sc}$ where $(0 \leq I_{sc} < 48)$.

Table 7.12   DCI Format N0 Information

Information	Size (Bits)	Meaning
Flag	1	If set to 1, indicates format N1 and if set to 0, indicates format N0
Subcarrier indication (I_{sc})	6	Determines the UL subcarriers allocated, n_{sc}
Resource assignment (I_{RU})	3	Determines number of uplink resource units, N_{RU}
Scheduling delay (I_{Delay})	2	The number of subframes elapsed from the end of NPDCCH subframe until the first UL slot of the NPUSCH
Modulation and coding scheme (I_{MCS})	4	Determines the modulation scheme and transport block size
Redundancy version (RV)	1	Indicates redundancy version of uplink transport block
Repetition number (I_{Rep})	3	Determines N_{Rep}
New data indicator	1	Indicates whether the transport block is a new one or a retransmission. This information is provided to the MAC sublayer
DCI subframe repetition number	2	Determines how many times the DCI (NPDCCH) is repeated
HARQ process number	1	Defines the HARQ process ID and only present if two HARQ processes are configured

Table 7.13   Repetition Number, I_{Rep}, for DCI Format N0

Repetition Number (I_{Rep})	Number of Repetition (N_{Rep})
0	1
1	2
2	4
3	8
4	16
5	32
6	64
7	128

Table 7.14 Resource Assignment, I_{RU}, for DCI Format N0

Resource Assignment (I_{RU})	Number of RU (N_{RU})
0	1
1	2
2	3
3	4
4	5
5	6
6	8
7	10

Table 7.15 Scheduling Delay, I_{Delay}, for DCI Format N0

Scheduling Delay (I_{Delay})	Number of Subframes (k)
0	8
1	16
2	32
3	64

Table 7.16 Subcarrier Indication, I_{sc}, on DCI Format N0 for Subcarrier Spacing of $\Delta f = 15$ KHz

Subcarrier Indication (I_{sc})	Set of Allocated Subcarriers (n_{sc})	Number of Subcarriers (N_{sc}^{UL})
0–11	I_{sc}	1
12–15	$3(I_{sc} - 12) + \{0, 1, 2\}$	3
16–17	$6(I_{sc} - 16) + \{0, 1, 2, 3, 4, 5\}$	6
18	$\{0, 1, 2, 3, 4, 5, 6, 7, 8, 9, 10, 11\}$	12

■ The total number of allocated subcarriers in a RU is computed according to n_{sc} and denoted N_{sc}^{RU}.

■ If $N_{sc}^{RU} > 1$, the constellation size is $Q_m = 2$ and $I_{TBS} = I_{MCS}$. If $N_{sc}^{RU} = 1$, I_{MCS} is used to determine the constellation size and transport block size index, I_{TBS}, according to Table 7.17.

■ Transport block size is determined using transport block size index, I_{TBS}, and resource assignment, I_{RU}, as shown in Table 7.18.

■ The UL transport block is transmitted and repeated over $N = N_{Rep} \times N_{RU} \times N_{slots}^{UL}$ slots. In a group of consecutive slots, a different RV of a block of the transport block is transmitted according to RV field in Table 7.12.

Table 7.17 I_{MCS} and I_{TBS} for DCI Format N0 when $N_{sc}^{RU} = 1$

Modulation and Coding Scheme (I_{MCS})	Constellation Size (Q_m)	TBS Index (I_{TBS})
0	1	0
1	1	2
2	2	1
3	2	3
4	2	4
5	2	5
6	2	6
7	2	7
8	2	8
9	2	9
10	2	10

Table 7.18 Transport Block Size (Bits) for NPUSCH

TBS Index (I_{TBS})	Resource Assignment (I_{RU})							
	0	1	2	3	4	5	6	7
0	16	32	56	88	120	152	208	256
1	24	56	88	144	176	208	256	344
2	32	72	144	176	208	256	328	424
3	40	104	176	208	256	328	440	568
4	56	120	208	256	328	408	552	680
5	72	144	224	328	424	504	680	872
6	88	176	256	392	504	600	808	1000
7	104	224	328	472	584	712	1000	1224
8	120	256	392	536	680	808	1096	1384
9	136	296	456	616	776	936	1256	1544
10	144	328	504	680	872	1000	1384	1736
11	176	376	584	776	1000	1192	1608	2024
12	208	440	680	1000	1128	1352	1800	2280
13	224	488	744	1128	1256	1544	2024	2536

■ The corresponding ACK/NACK response for UL transmission is signalled implicitly by the new data indicator (NDI) field. If NDI is toggled, it indicates an ACK for the previous UL transmission.

7.10.9.12 DCI Format N0 Example

To provide a numeric example of how DCI format N0 is decoded by the NB-IoT device, consider Table 7.19.

Table 7.19   DCI Format N0 Fields

Field	Size (Bits)	Value
Flag	1	0
Subcarrier indication (I_{sc})	6	12
Resource assignment (I_{RU})	3	1
Scheduling delay (I_{Delay})	2	1
Modulation and coding scheme (I_{MCS})	4	1
Redundancy version (RV)	1	0
Repetition number (I_{Rep})	3	1

DCI information received in Table 7.19 indicates the following:

■ UE received a random access response messages (RAR), as in Table 6.4, indicating a subcarrier spacing of $\Delta f = 15$ KHz.

■ Flag sets to 0 means this is DCI Format N0.

■ Subcarrier indication, I_{sc}, set to 12, indicates allocated set of subacarrier is {0,1,2} as in Table 7.16.

■ Resource assignment, I_{RU}, set to 1, indicates that N_{RU} is 2 as in Table 7.14.

■ Scheduling delay, I_{Delay}, of 1, indicates that after the NPDCCH ends in DL subframe n, NPUSCH subframes starts $k = 16$ DL subframe after DL subframe n and repeated N consecutive subframes as in Tables 7.12 and 7.15.

■ Modulation and coding scheme, I_{MCS}, value of 1, indicates that $I_{MCS} = I_{TBS}$ as $N_{SC}^{RU} > 1$.

■ Repetition number, I_{Rep}, set to 2, indicates that N_{Rep} is 4 as in Table 7.13.

From the above, UE infers the following:

■ n_{sc} is a set of allocated carrier {0,1,2} and the total number of allocated subcarriers in a RU, N_{sc}^{RU}, is 3. According to Table 7.29, N_{Slots}^{UL} is 8.

■ The constellation size for QPSK is $Q_m = 2$ since $N_{sc}^{RU} = 3 > 1$.

■ The transport block size is 56 bits as in Table 7.18 since $I_{TBS} = 1$ and $I_{RU} = 1$.

7.10.9.13   DCI Format N1

This format is used to indicate a downlink grant in the NPDSCH for a single downlink carrier. It is also used to indicate a random access procedure triggered by an NPDCCH order (e.g., triggered by eNodeB) or for a notification of SC-MCCH change. The CRC of this format is scrambled with the C-RNTI, temporary C-RNTI, RA-RNTI, SI-RNTI, or G-RNTI. The information carried in DCI format N1 is shown in Table 7.20. The maximum total size of DCI format N1 is 24 bits.

If NPDCCH order indicator is set to 1, Starting number of NPRACH repetitions, I_{Rep}, Subcarrier indication of NPRACH, I_{sc}, and Carrier indication of NPRACH are only used and all other remaining information are set to zero. When NPDCCH order indicator is set to 1, UE transmits a random access preamble on the non-anchor carrier indicated by Carrier indication of NPRACH and on coverage enhancement level mapped to NRACH resource with number of preamble repetitions, I_{Rep}. If Carrier indication field is zero, UE transmits the random access preamble on the anchor carrier. Random access preamble is used with $\Delta f = 3.75$ KHz and the random access preamble to be used is indicated by the subcarrier indication field, I_{sc}, where $0 \le I_{sc} < 48$.

NPDSCH is located in DL subframes that are designated as NB-IoT DL subframes. These subframes are announced to UE through a SIB1-NB or RRConnectionSetup PDU, in subframePattern10 or subframePattern40, as in Table 7.1, excluding those frames used by SIB1-NB, NPBCH, NPSS, or NSSS.

NPDSCH transmission can be delayed a number of subframes k where $0 \le k < 128$. That is, after subframe $n + 5$, NPDSCH subframes starts k subframes after subframe $n + 5$ and repeats N consecutive subframes. When the UE detects format N1 or N2 that ends in subframe, n, NPDSCH starts in subframe $n + 5 + k$. NPDSCH is also repeated in N consecutive subframes starting from subframe $n + 5 + k$. k is determined by the scheduling delay field, I_{Delay}, as in Tables 7.20 and 7.21. $k = 0$ for DCI format N2. In other words:

- NPDCCH transmission ended in subframe n.

- NPDSCH is repeated in a number of consecutive subframes where the transport block and its repetitions are spread across $N = N_{Rep} \times N_{SF}$ (both are determined according to Tables 7.20, 7.22, and 7.23).

- The modulation scheme for NPDSCH is QPSK ($Q_m = 2$). The transport block size is determined from modulation and coding scheme, I_{MCS}, by setting $I_{TBS} = I_{MCS}$. Both I_{TBS} and resource assignment field, I_{SF}, determines the transport block size as in Table 7.24. Transport block is spread across N_{SF} subframes and each transport block is repeated N_{Rep} subframes.

Table 7.20   DCI Format N1 Fields

Field	Size (Bits)	Meaning
Flag	1	If set to 1, indicates format N1 and if set to 0, indicates format N0
NPDCCH order indicator	1	If set to 1, indicates a random access procedure initiated by an NPDCCH order
SC-MCCH change notification	2	Indicates a change in SC-MCCH
Starting number of NPRACH repetitions (I_{Rep})	2	Determines repetition number, N_{Rep}, for NPRACH
Subcarrier indication of NPRACH (I_{sc})	6	Used to allocate a subcarrier for NPRACH, $n_{sc} = I_{sc}$
Carrier indication of NPRACH	4	Indicates the carrier index, in ul-ConfigList, used to transmit random access preamble. This field is used for NPDCCH order and if there is non-anchor carrier configured
Scheduling delay (I_{Delay})	3	Number of subframes elapsed from the end of NPDCCH plus 5 subframes until the first subframe of the PDSCH
Resource assignment (I_{SF})	3	Determines, N_{SF}, and Transport block size
Modulation and coding scheme (I_{MCS})	4	Determines $I_{TBS} = I_{MCS}$
Repetition number (I_{Rep})	4	Determines N_{Rep}
New data indicator	1	Indicates whether the transport block is a new one or a retransmission. This information is provided to the MAC sublayer
HARQ-ACK resource	4	Indicates the subcarrier allocated for ACK/NACK
DCI subframe repetition number	2	Determines how many times the DCI (NPDCCH) is repeated
HARQ process number	1	Defines the HARQ process ID and only present if two HARQ processes are configured

Table 7.21 Scheduling Delay, I_{Delay}, for DCI Format N1

Scheduling delay (I_{Delay})	Number of Subframes (k)
0	0
1	4
2	8
3	12
4	16
5	32
6	64
7	128

Table 7.22 Number of Subframes, N_{SF}, for DCI Format N1

Resource Assignment (I_{SF})	Number of Subframes (N_{SF})
0	1
1	2
2	3
3	4
4	5
5	6
6	8
7	10

System information, SIB1-NB and other SIBs, are carried on NPDSCH. SIB1-NB has a special transmission schedule as explained in Section 3.7.2. The subframes used for SIB1-NB transmission are known to the UE according to Section 3.7.2. If the subframe carries a SIB1-NB, N_{Rep} is set according to the value of the parameter schedulingInfoSIB1 as shown in Table 3.7. Transport block size, I_{TBS}, for SIB1-NB is determined according to Table 7.25.

After the NPDSCH transmission is completed in subframe m, the UE starts to transmit ACK/NACK to eNodeB. The UE starts sending ACK/NACK k subframes after the end of NPDSCH. That is, after DL subframe $(m + k - 1)$ ends, UE starts to transmit ACK/NACK response. The UL subcarrier and k is determined from HARQ-ACK resource field in Tables 7.20 and 7.26. ACK/NACK uses only a single subcarrier irrespective of the subcarrier spacing.

The ACK/NACK response is transmitted using NPUSCH format 2 and in N consecutive UL slots. $N = N_{Rep}^{ACK} \times N_{slots}^{UL}$, where the value

Table 7.23 Number of Repetition Subframes, N_{Rep}, for DCI Format N1

Repetition Number (I_{Rep})	Number of Subframes (N_{Rep})
0	1
1	2
2	4
3	8
4	16
5	32
6	64
7	128
8	192
9	256
10	384
11	512
12	768
13	1024
14	1536
15	2048

Table 7.24 Transport Block Size (Bits) for NPDSCH

TBS Index (I_{TBS})	Resource Assignment (I_{SF})							
	0	1	2	3	4	5	6	7
0	16	32	56	88	120	152	208	256
1	24	56	88	144	176	208	256	344
2	32	72	144	176	208	256	328	424
3	40	104	176	208	256	328	440	568
4	56	120	208	256	328	408	552	680
5	72	144	224	328	424	504	680	872
6	88	176	256	392	504	600	808	1032
7	104	224	328	472	584	680	968	1224
8	120	256	392	536	680	808	1096	1352
9	136	296	456	616	776	936	1256	1544
10	144	328	504	680	872	1032	1384	1736
11	176	376	584	776	1000	1192	1608	2024
12	208	440	680	904	1128	1352	1800	2280
13	224	488	744	1128	1256	1544	2024	2536

Table 7.25   Transport Block Size (Bits) for NPDSCH Carrying SIB1-NB

I_{TBS}	0	1	2	3	4	5	6	7	8	9	10	11
TBS (Bits)	208	208	208	328	328	328	440	440	440	680	680	680

Table 7.26   ACK/NACK Resource Field when $\Delta f = 15$ KHz

HARQ-ACK Resource	ACK/NACK Subcarrier	k
0	0	13
1	1	13
2	2	13
3	3	13
4	0	15
5	1	15
6	2	15
7	3	15
8	0	17
9	1	17
10	2	17
11	3	17
12	0	18
13	1	18
14	2	18
15	3	18

of N_{Rep}^{ACK} is given by the RRC parameter ack-NACK-NumRepetitions (Table 7.1) and N_{slots}^{UL} is the number of slots of the RU as in Table 7.29.

7.10.9.14   DCI Format N2

This format is used to indicate paging, direct indication (indication of a change in SIBs contents), downlink SC-MCCH, or a notification of SC-MCCH change. The information carried in DCI format N2 is shown in Table 7.27. The total size of DCI format N2 is 15 bits (Direct Indication information field is only used if the Flag field is set to 0 and NPDCCH is scrambled by P-RNTI).

The interpretation of these fields is the same as that for DCI format N1. DCI format N2 is encoded by the UE to extract Paging PDU on NPDSCH the same way as explained in DCI format N1.

The UE distinguishes the different formats of the DCI by checking the CRC. If the CRC is scrambled with the RA-RNTI, the NPDCCH content is a DCI format N1 and includes NPDSCH resource for the

Table 7.27 DCI Format N2 Fields

Field	Size (Bits)	Meaning
Flag	1	If set to 1, indicates Paging and if set to 0, indicates direct indication
SC-MCCH change notification	1	Indicate a change in SC-MCCH
Direct Indication information	8	Indicates system information update (without the need to receive a Paging PDU).
Resource assignment (I_{SF})	3	Determines N_{SF} and Transport block size
Modulation and coding scheme (I_{MCS})	4	Determines $I_{TBS} = I_{MCS}$
Repetition number (I_{Rep})	4	Determines N_{Rep}
DCI subframe repetition number	3	Determines how many times the DCI (NPDCCH) is repeated

MAC PDU containing the random access response message as explained in Section 6.3.2.

If the CRC is scrambled with the temporary C-RNTI, it indicates DCI format N1 which includes NPDSCH that contains Msg4 as explained in Section 6.3.3.

If the CRC is scrambled with the C-RNTI, the first bit in the message indicates whether it is a DCI format N0 or N1. If format N0, it indicates NPUSCH transmission opportunity while if format N1, it indicates NPDSCH transmission.

If the CRC is scrambled with the SI-RNTI, it indicates DCI format N1 which includes NPDSCH that contains SIB1-NB or other SIBs.

If the CRC is scrambled with G-RNTI, it is a DCI format N1. If the CRC is scrambled by P-RNTI or SC-RNTI, this indicates DCI format N2.

7.10.9.15 Spatial Multiplexing and Transmit Diversity

There is no support for spatial multiplexing while transmit diversity is supported up to two layers. Two antenna ports are supported. Transmit diversity can be used for NPDSCH and NPBCH to increase the reliability of message reception by the UE. Transmit diversity is illustrated in Section 7.10.8.7.

7.11   Uplink Physical Channels and Structure

The uplink channels consist of the following physical channels:

■ Narrowband Physical Uplink Shared Channel, NPUSCH.

■ Narrowband Physical Random Access Channel, NPRACH.

In addition to the following physical signals:

■ Narrowband demodulation reference signal.

UE does not have an NPUCCH.

7.11.1   *Uplink transmission scheme SC-FDMA*

Uplink transmission scheme is based on a single-carrier FDMA (SC-FDMA). SC-FDMA is also described as DFT-spread OFDM (DFTS-OFDM) which performs DFT precoding before the uplink OFDM modulator. There is only one cyclic-prefix length defined which corresponds to seven OFDM symbols per slot, respectively.

For $\Delta f = 15$ KHz subcarrier spacing,

■ Normal cyclic prefix: $T_{CP} = 5.2$ us (OFDM symbol #0), $T_{CP} = 4.7$ us (OFDM symbol #1 to #6).

■ $T_s = 1/(2048 \times \Delta f)$.

For $\Delta f = 3.75$ KHz subcarrier spacing,

■ Normal cyclic prefix: $T_{CP} = 8.3$ us (OFDM symbol #0 to #6).

■ $T_s = 1/(8192 \times \Delta f)$, where T_s is a sampling time for an OFDM symbol.

Both single-tone transmission and multitone transmission are used by UE. For single-tone transmission, there are two numerologies defined: 3.75 and 15 KHz subcarrier spacing.

In the frequency domain, resource blocks are not defined but instead a RU is defined. If the uplink subcarrier spacing $\Delta f = 15$ KHz, there are 12 consecutive subcarriers. If the uplink subcarrier spacing $\Delta f = 3.75$ KHz, there are 48 consecutive subcarriers. Table 7.28 summarizes the uplink slot configuration.

Figures 7.22 and 7.23 show the uplink slot duration and number of subcarriers for both subcarrier spacing of 15 and 3.75 KHz, respectively.

Single-tone transmission with 3.75 KHz subcarrier spacing is organized into slots with 2 ms duration, each of which consists of seven symbols located from beginning of the slot. The slot boundary is aligned with subframe boundaries of frame structure Type 1. One symbol of

Table 7.28   Uplink Slot Configuration

Subcarrier Spacing (KHz)	Maximum Number of Subcarriers	SC-FDMA Symbols	Slot Duration (ms)
$\Delta f = 3.75$	48	7	2
$\Delta f = 15$	12	7	0.5

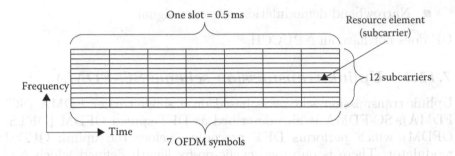

Figure 7.22: Uplink slot for $\Delta f = 15$ KHz.

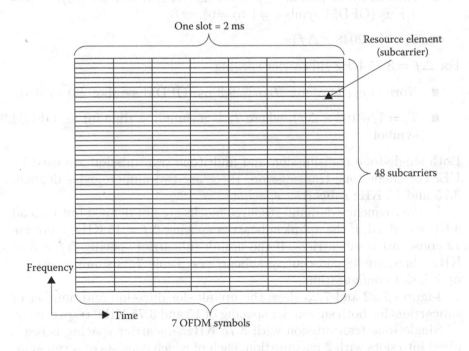

Figure 7.23: Uplink slot for $\Delta f = 3.75$ KHz.

3.75 KHz subcarrier spacing consists of 275 us of symbol duration (including the cyclic-prefix of 8.3 us). Since the 2 ms slot has seven symbols, a remaining, 75 us, of the slot is empty and used as a guard period.

For multitone transmission, there are 12 consecutive uplink subcarriers with uplink subcarrier spacing of $\Delta f = 15$ KHz. The subcarriers can be grouped into sets of 3, 6, or 12 consecutive subcarriers.

A RU, schedulable for single-tone NPUSCH with UL-SCH transmission, is defined as a single 3.75 KHz subcarrier for 16 slots (32 ms) or a single 15 KHz subcarrier for 16 slots (8 ms).

A RU, schedulable for multitone NPUSCH with UL-SCH transmission is defined as 3 subcarriers for 4 ms; or 6 subcarriers for 2 ms; or 12 subcarriers for 1 ms.

A RU, schedulable for NPUSCH with ACK/NACK transmission uses NPUSCH format 2. It is defined as a single 3.75 KHz subcarrier for 8 ms or a single 15 KHz subcarrier for 2 ms.

An NPUSCH (UL-SCH) transport block can be scheduled over one or more than one RU in time.

Table 7.29 summarizes the number of used subcarriers, number of slots, subcarrier spacing used for both subcarrier spacing. Figure 7.24 shows a RU consists of a single subcarrier spanning 16 slots (a slot is 2 ms for $\Delta f = 3.75$ KHz and 0.5 ms for $\Delta f = 15$ KHz). Figure 7.25 shows a RU consists of six subcarriers spanning four slots.

7.11.2   Resource grid

Resource grid is different from the resource block used in the downlink. An uplink slot takes the shape as in Figure 7.22 or 7.23 for $\Delta f = 15$ KHz or $\Delta f = 3.75$ KHz, respectively. It consists of seven SC-FDMA symbols in time domain and a number of subcarriers in frequency domain.

Table 7.29   NPUSCH Format

NPUSCH Format	Subcarrier Spacing (KHz)	Number of Subcarriers	Number of Slots	Total slots Duration (ms)	Number of SC-FDMA symbols per slot
1	3.75	1	16	32	7
	15	1	16	8	
		3	8	4	
		6	4	2	
		12	2	1	
2	3.75	1	4	8	
	15	1	4	2	

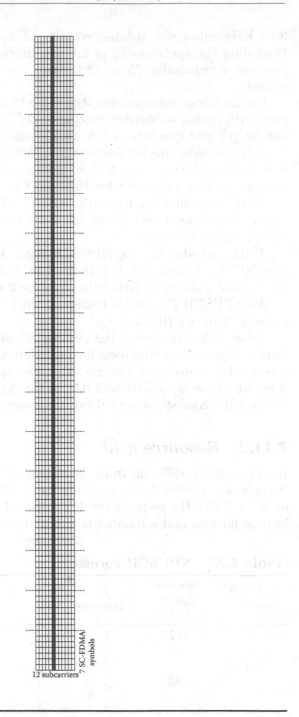

12 subcarriers

7 SC-FDMA symbols

Figure 7.24: Single-tone RU for 16 slots for $\Delta f = 15$ KHz as in Table 7.29 (total duration of 8 ms).

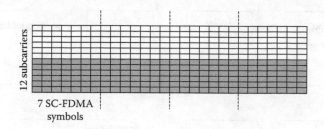

Figure 7.25: Multitone (six subcarriers) RU for four slots for $\Delta f = 15$ KHz as in Table 7.29 (total duration of 2 ms).

7.11.3 NPUSCH

NPUSCH is used to transmit uplink transport block. A maximum of only one transport block is transmitted per carrier. When the MAC sublayer passes a transport block (or MAC PDU) to PHY layer for uplink transmission, NPUSCH performs the following functionalities:

- CRC insertion: 24 bit CRC.

- Channel coding: Turbo coding.

- Rate matching.

- Physical-layer hybrid-ARQ processing.

- Scrambling: UE-specific scrambling.

- Modulation: $\pi/2$-BPSK and $\pi/4$-QPSK for single-tone transmission of NPUSCH, and QPSK for multitone transmission of NPUSCH.

- Mapping to assigned resources.

Figure 7.26 shows the different functionalities for processing uplink channels.

7.11.3.1 CRC Calculation

As in Section 7.10.8.1 and the same for NPDSCH, the same CRC is used.

7.11.3.2 Turbo Coding

The turbo encoder is a Parallel Concatenated Convolutional Code (PCCC) with two eight-state constituent encoders and one turbo code

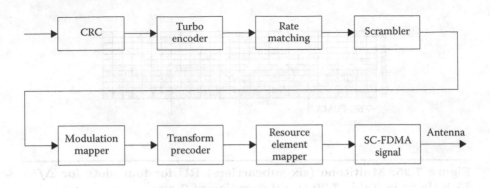

Figure 7.26: Uplink channel processing.

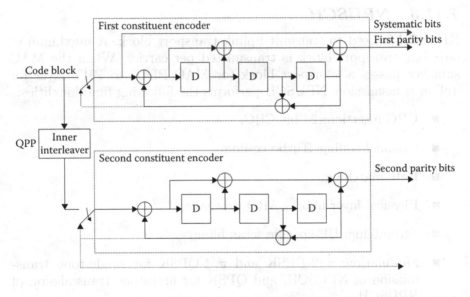

Figure 7.27: Turbo encoder with rate 1/3.

internal interleaver. The coding rate of turbo encoder is 1/3. The turbo encoder is shown in Figure 7.27.

The transfer function of the PCCC is:

$$G(X) = \left[1, \frac{1 + X^2 + X^3}{1 + X + X^3}\right].$$

The input bits to the turbo encoder are $b_0, b_1, b_2, \ldots, b_{B-1}$ where B are the transport block bits and CRC parity bits. The output from the turbo encoder is s_k^0, p_k^1, p_k^2 where s is the systematic bits, p_k^1 and p_k^2 are the first and second parity bits, $k = 0, 1, 2, \ldots, K - 1$, and $K = B + 4$.

The shift registers of the turbo coder have initial values of zeros when starting to encode the input bits.

7.11.3.3 Rate Matching

The rate matching block is similar to the one used for NPDSCH in Section 7.10.8.3 except that the subblock interleaver and bit collection and pruning uses different parameters.

The output from the turbo encoder is provided as an input to the rate matching block shown in Figure 7.29. The three information bit streams, s_k^0, p_k^1, p_k^2, obtained from the turbo encoder are provided as an input to each subblock interleaver which interleaves them separately. The interleaved bits are inserted into a circular buffer with the systematic bits inserted first and followed by alternating insertion of the second and third parity bits as in Figure 7.28. The bit selection extracts consecutive bits from the circular buffer to an extent that matches the number of available resource elements in the resource blocks (i.e., soft buffer size) assigned for the transmission. The exact set of bits to be extracted depends on the RV corresponding to different starting points for the extraction of coded bits from the circular buffer. There are four different alternatives for the RV. The output sequence from the bit selection is denoted by $e_0, e_1, e_2, \ldots, e_{E-1}$, where E is the number of rate matched bits. Not all RV are used for NPUSCH and only RV = 0 or RV = 2 are used.

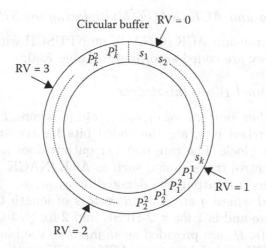

Figure 7.28: Circular buffer with redundancy version (RV).

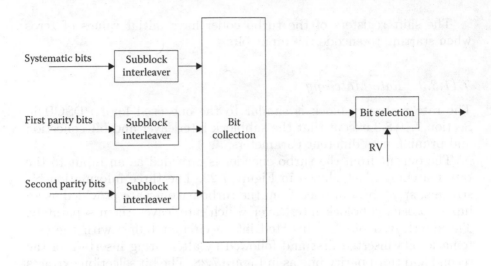

Figure 7.29: Rate matching for turbo encoder with rate 1/3.

Table 7.30 ACK/NACK Channel Coding

ACK/NACK	ACK/NACK Codeword $(b_0, b_1, b_2, \ldots, b_{15})$
0	0,0,0,0,0,0,0,0,0,0,0,0,0,0,0,0
1	1,1,1,1,1,1,1,1,1,1,1,1,1,1,1,1

7.11.3.4 Data and ACK/NACK Multiplexing on NPUSCH

If the UE is to transmit ACK or NACK on NPUSCH without any data on UL-SCH, they are coded according to Table 7.30.

7.11.3.5 Channel Block Interleaver

The rate matcher output, $e_0, e_1, e_2, \ldots, e_{E-1}$, where E is the number of rate matched bits, are the coded bits for transmission of the given transport block. The rate matcher output does not contain the bits used for control transmission such as ACK/NACK. Rate matcher bits are reordered as output bits denoted by $g_0, g_1, g_2, \ldots, g_{H-1}$, where $H = E/Q_m$ and where g are column vectors of length Q_m. Q_m is the constellation size and is 1 for $\pi/2$-BPSK and 2 for $\pi/4$-QPSK.

The data bits, H, are provided as an input to the block interleaver. The ACK/NACK coded bits (i.e., ACK/NACK codeword) are combined with the interleaved data bits from the block interleaver to provide the bit sequence denoted by $h_0, h_1, h_2, \ldots, h_{H-1}$.

7.11.3.6   Scrambling

The codeword, $h_0, h_1, h_2, \ldots, h_{H-1}$, where H is the number of bits transmitted in the codeword is scrambled before modulated. Codeword is multiplied, as an X-OR operation, by a UE-specific scrambling sequence yielding the scrambling output, $\tilde{h}_0, \tilde{h}_1, \tilde{h}_2, \ldots, \tilde{h}_{H-1}$.

7.11.3.7   Modulation

Each scrambled codeword, $\tilde{h}_0, \tilde{h}_1, \tilde{h}_2, \ldots, \tilde{h}_{H-1}$, is modulated using either BPSK or QPSK which corresponds to either one bit or two bits per complex-value symbol.

There are two configurations (or format) for NPUSCH as in Table 7.29. PUSCH format 1 is used for data transmission on UL-SCH. NPUSCH format 2 is used for uplink control information (e.g., HARQ ACK/NACK transmissions). Format 2 always has a RU of one subcarrier irrespective of subcarrier spacing. ACK/NACK corresponding to NPDSCH is transmitted with single-tone transmission on NPUSCH, with frequency resource and time resource indicated by downlink grant.

For NPUSCH format 2, the modulation scheme is always $\pi/2$-BPSK. For NPUSCH format 1, if the RU is one subcarrier, $\pi/2$-BPSK or $\pi/4$-QPSK can be used. All other RUs in format 1 uses QPSK.

Table 7.31 shows the exact modulation format to be used where $\pi/4$-QPSK and $\pi/2$-BPSK can be used. Constellation mapping for these

Table 7.31   NPUSCH Modulation

Transport Channel	Modulation
Format 1 with single subcarrier	$\pi/2$-BPSK, $\pi/4$-QPSK
Format 1 with multiple subcarriers	QPSK
Format 2 with single subcarrier	$\pi/2$-BPSK

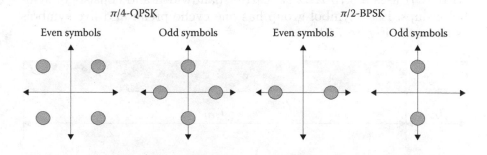

Figure 7.30: $\pi/2$-BPSK and $\pi/4$-QSPK constellation mapping.

modulation schemes are shown in Figure 7.30. $\pi/4$-QPSK is the same as QPSK but the constellation is shifted an angle of $\pi/4$ for odd-numbered symbols. Also, $\pi/2$-BPSK is the same as BPSK but constellation is shifted an angle of $\pi/2$ for odd-numbered symbols.

7.11.3.8 Layer Mapping

UE supports only one layer for uplink. After modulation, the modulation symbols for the codeword are mapped to one layer. Complex-valued modulation symbols, $m_0, m_1, m_2, \ldots, m_{N_{sym}-1}$, for the codeword is mapped into one layer $x_i^0 = m_i$, where $i = 0, 1, 2, \ldots, N_{sym} - 1$.

7.11.3.9 Transform Precoding

The number of symbols, N_{sym}, are divided into a number of sets, each set consists of N_{sc}^{RU} modulation symbols and corresponds to one SC-FDMA symbol. N_{sc}^{RU} is the number of subcarriers as in Table 7.16.

Since there is only a single antenna port for the uplink, the modulation symbols are mapped into resource elements directly without any precoding.

7.11.3.10 Mapping to Physical Resources

One or more RUs, N_{RU}, can be assigned to the UE for its uplink transmission on NPUSCH. The number of complex-value symbols, N_{sym}, are mapped to each RU starting with subcarriers first and then to each SC-FDMA symbol. The symbols are mapped to N_{slots}^{UL} slots for each RU and then repeated a number of times N_{Rep}. The mapping continues to the remaining RUs.

7.11.4 NPRACH

The NPRACH is used for random access preamble transmission. NPRACH uses a 3.75 KHz subcarrier bandwidth and consists of symbol groups. Each symbol group has one cyclic prefix and five symbols as shown in Figure 7.31.

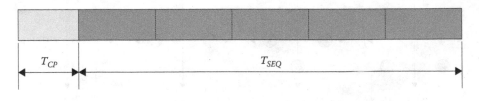

Figure 7.31: Random access symbol group.

Table 7.32   Symbol Group Parameters

Preamble Format	T_{CP} (us)	T_{SEQ} (ms)
0	66	1.333
1	266	1.333

Two preamble formats are defined, format 0 and format 1 as shown in Table 7.32 suitable for different maximum cell sizes. The five symbols have a single CP of $T_{CP} = 66$ us for format 1 and $T_{CP} = 266$ us for format 2. The duration of a symbol for both formats are $T_{SEQ} = 1.333$ ms. The symbol group is thus 1.4 and 1.6 ms for format 1 and 2, respectively. The preamble format to be used is broadcasted in the system information.

The long CP of $T_{CP} = 266$ us can be used for large cells with radius in the range of 40 km, and the short CP of $T_{CP} = 66$ us can be used for cells with radius in the range of 10 km.

The preamble is composed of four symbol groups transmitted without gaps. Each symbol group is transmitted on a single subcarrier which are used in a frequency hopping manner. Each subcarrier in a symbol group hops by one or six subcarriers in frequency. Frequency hopping is restricted to a contiguous set of 12 subcarriers. The preamble transmission can be repeated multiple times, using the same transmission power on each repetition. Figure 7.32 shows the frequency hopping pattern of fours symbol groups in a random access preamble.

The procedure and parameters used for random access are described in Section 6.3.2. After the first subcarrier is selected for the transmission of the first preamble symbol group, the next three subcarriers for the next three symbol groups are determined by a frequency hopping criteria which depends only on the location of the first subcarrier. When the symbol groups are repeated, the first subcarrier is selected according to a pseudo-random hopping criteria where physical cell ID, PHY_{CELL}^{ID}, is used as an input.

This frequency hopping criteria guarantees that subcarriers selection results in a frequency hopping scheme which can accommodate collision-free transmissions from UE as capacity permits. The preamble sequence is built upon a Zadoff–Chu sequence which depends on the subcarrier location. Performance study of the random access procedure and its possible optimization have been studied in [36]–[38].

7.11.5   *Demodulation reference signal*

Demodulation Reference Signal (DMRS) is transmitted from the UE to eNodeB. It is transmitted in the same RUs used for NPUSCH. DMRS

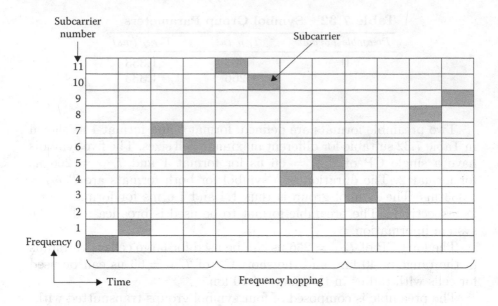

Figure 7.32: Frequency hopping of four symbol groups of a random access preamble.

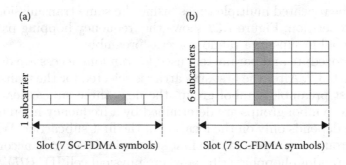

Figure 7.33: DMRS for NPUSCH format 1 when subcarrier spacing is (a) 3.75 KHz and (b) 15 KHz.

is transmitted on either 1 or 3 SC-FDMA symbols. For NPUSCH format 1, SC-FDMA symbol #4 and #3 are used for subcarrier spacing of 3.75 or 15 KHz, respectively, as shown in Figure 7.33. For NPUSCH format 2, and in case of subcarrier spacing of 3.75 KHz, SC-FDMA symbol #0, #1, #2 are used while for subcarrier spacing of 15 KHz, SC-FDMA symbol #2, #3, #4 are used. NPUSCH format 2 is shown Figure 7.34. Other subcarriers not used by DMRS are used for NPUSCH.

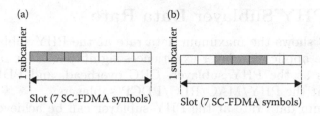

Figure 7.34: DMRS for NPUSCH format 2 when subcarrier spacing is (a) 3.75 KHz and (b) 15 KHz.

7.11.6 *Uplink power control*

The UE transmit power for NPUSCH is based on whether the number of repetitions of NPUSCH is less than or greater than two repetitions. If the NPUSCH repetitions is two or less than two, UE uses this equation to calculate the NPUSCH power in a slot n for a cell:

$$P(n) = \min \left[\begin{array}{c} P_{MAX}(n) \\ 10\log_{10}(M(n)) + P_O(j) + \alpha(j).P_L \end{array} \right] \text{ dBm,}$$

Otherwise, if the number of repetitions is greater than two, UE uses this equation

$$P(n) = P_{MAX}(n) \text{ dBm,}$$

where

- $P_{MAX}(i)$ is a cell-specific maximum transmit power in slot n.

- $M(n)$ equals to {1/4} for 3.75 KHz subcarrier spacing and a single value from the set of {1, 3, 6, 12} for 15 KHz subcarrier spacing.

- P_O is a combination of different parameters signalled by the RRC, that depends on whether the transport block is for UL-SCH data ($j = 1$) or for the RACH grant message ($j = 2$).

- $\alpha(j)$ is a path-loss factor that is provided by RRC for NPUSCH format 1; otherwise, it has a fixed value of 1.

- P_L is the path loss that is measured and estimated by the UE in dB. The path-loss factor is used to indicate how strong the path loss shall be compensated.

7.12   PHY Sublayer Data Rate

Table 7.33 shows the maximum data rate at the PHY sublayer. Data rate at the application layer experiences significant lower rate due to repetitions at the PHY sublayer, CRC overhead, and PDU headers overhead at the PHY/MAC/RLC/PDCP sublayers.

Maximum data rate at the PHY sublayer can be achieved if there are no repetitions. In the downlink, one transport block transmission in a subframe yields the maximum downlink data rate. Similarly, for the uplink, one transport block transmission within the number of slots configuration yields the maximum uplink data rate.

For the downlink and as in Table 7.24, a single transport block size of 1544 bits can be achieved when the resource assignment, $I_{SF} = 5$. $I_{SF} = 5$ indicates that the number of subframes, N_{SF}, equals 6 (6 ms) for transmission of this transport block size (Table 7.22). The number of repetition subfarmes, N_{Rep}, equals 1 which means that the transport block is transmitted only once and with no other repetition.

For the uplink and when $\Delta f = 15$ KHz, Table 7.18 indicates that a single transport block size of 1544 bits can be achieved when the resource assignment, $I_{RU} = 5$. $I_{RU} = 5$ indicates that the number of RUs, N_{RU}, equals 6 (6 ms) for transmission of this transport block size (Table 7.14). The number of repetition subfarmes, N_{Rep} equals 1 which means that the transport block is transmitted only once and with no other repetition. Maximum uplink data rate for $\Delta f = 15$ KHz is also achieved with a configuration of 12 subcarriers and 2 slots as shown in Table 7.29.

For the uplink and when $\Delta f = 3.75$ KHz, Table 7.18 indicates that a single transport block size of 872 bits can be achieved corresponding to a TBS index, $I_{TBS} = 10$ and a resource assignment, I_{RU}, equals 4. This is because with $\Delta f = 3.75$ KHz, the number of subcarriers is 1 and the maximum TBS index, I_{TBS}, is 10 (Table 7.17). $I_{RU} = 4$ indicates that the number of RUs, N_{RU}, equals 5 for transmission of this transport block size (Table 7.14). When $\Delta f = 3.75$ KHz, the uplink slot uses a configuration of 1 subcarrier, 16 slots, and each slot is 2 ms, as shown in Table 7.29, thus yielding a $16 \times 2 \times 5 = 160$ ms duration of a transport block size of 872 bits.

Table 7.33   Maximum PHY Data Rate for Downlink and Uplink

NPDSCH Data Rate (Kbps)	NPUSCH Data Rate for $\Delta f = 15$ KHz (Kbps)	NPUSCH Data Rate for $\Delta f = 3.75$ KHz (Kbps)
257.3	257.3	5.5

It is worth noting that, as in Table 3.29, the maximum transport block size for a Cat-NB1 UE is 680 bits and 1000 bits for the downlink and uplink, respectively. The maximum transport block size for a Cat-NB2 UE is 2536 bits for either the downlink or uplink. However, the maximum PHY data rate is similar for both Cat-NB1 and Cat-NB2 although Cat-NB2 UE has a larger transport block size than Cat-NB1 UE. This is because the number of downlink subframes (or uplink RUs) needed for such a larger transport block size also increases in time thus limiting the increase in the maximum PHY data rate.

Chapter 8

Quality of Service Architecture

8.1 NB-IoT Quality of Service

A bearer is the level of granularity that is used for transferring packets and traffic end-to-end. This means that any packet or traffic mapped to the same bearer receives the same treatment (e.g., same packet loss rate, priority).

A radio bearer is the bearer established on the wireless channel between the UE and eNodeB. Additional bearers are established in the network so that the UE can have a connectivity to the Internet, applications, and services. In the network, additional two bearers are established: an S1 bearer between eNodeB and S-GW and an S5/S8 bearer between the S-GW and PDN gateway (P-GW). All three bearers are concatenated together to establish a single end-to-end bearer, called EPS bearer, and to provide connectivity to UE [39, 40].

EPS bearer connects the UE to the Internet, applications, or other services through the P-GW and PDN. PDN provides the UE with connectivity to the Internet. The UE typically has one connection to a P-GW which is associated with a single PDN. The PDN is represented by an Access Point Name (APN) which is a name, according to DNS naming convention, describing the PDN.

The UE and core network establish a first EPS bearer to connect the UE to the PDN which remains established during the lifetime of the connection to this PDN. This EPS bearer provides an always-on

IP connectivity to the UE. This bearer is called the default bearer (a dedicated bearer is an additional bearer that is established to the same PDN). NB-IoT UE supports only default bearer and does not support the establishment of a dedicated bearer. The PDN uses the RAT Type (RAT Type = NB-IoT) to ensure that only default EPS bearer and no dedicated EPS bearer are activated by NB-IoT UE.

Figure 8.1 shows the end-to-end EPS bearer comprising intermediate bearers between the NB-IoT device and the network components, eNodeB, S-GW, and P-GW. The figure also illustrates the delay incurred between a UE transmitting traffic to another UE through the cellular infrastructure.

The network has full control of the EPS bearer. It can additionally assign or modify the QoS parameters values to the EPS bearer. The default bearer has initial QoS parameters that are assigned by the network based on the UE subscription information stored in the HSS. The network defines two type of EPS bearers: Guaranteed Bit Rate (GBR) and Non-Guaranteed Bit Rate (Non-GBR) bearers.

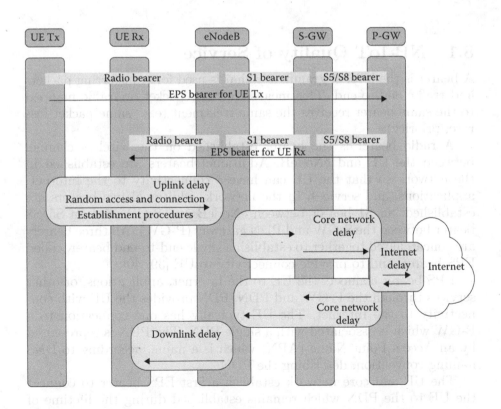

Figure 8.1: End-to-end EPS bearer and delay.

The default EPS bearer is always a non-GBR bearer. That is, the network does not guarantee bit rate to this bearer and a non-GBR bearer is expected to receive congestion-related packet drops.

The QoS parameters defined for Non-GBR bearer in each DL or UL direction are:

- QoS Class Identifier (QCI): This is a scalar value assigned by the network that defines the traffic characteristics. Each QCI value is mapped to a standardized number of QoS characteristics. This parameter is set by the network based on the USIM subscription.

- Allocation and Retention Priority (ARP): This defines the priority of EPS bearer which is used by the network in case it needs to drop or modify a bearer if there is a resource congestion or starvation. This parameter is configured and set by the network.

The QCI and ARP are not signalled by any network component. The network needs to guarantee the QCI characteristics when configuring the network and its components.

A UE that needs to use two DRBs is one use case where it can be beneficial to use two EPS bearers with different ARP values. The network can map the first DRB traffic to one bearer with a higher ARP value, and the second DRB traffic to another bearer with a lower ARP value. In a congestion situation, the eNodeB can drop the EPS bearer of the DRB with lower ARP value without affecting the other DRB traffic.

In addition, for each UE, the network maintains the following parameters for each DL or UL direction:

- APN Aggregate Maximum Bit Rate (APN-AMBR): This is the aggregate maximum bit rate transmitted or received by each APN accessed by UE. This is the sum of all traffic across DRBs and/or SRBs.

- UE Aggregate Maximum Bit Rate (UE-AMBR): This is the aggregate maximum bit rate transmitted or received by a UE.

The network assigns the QoS parameters to the default EPS bearer based on the UE USIM subscription. USIM of the UE contains all QoS subscription information. When the UE is attached to the network, the network retrieves QoS information stored for this UE from the HSS. HSS stores the QoS parameters, QCI, ARP, APN-AMBR, and UE-AMBR. A sample of the USIM subscription information stored in HSS is shown in Table 8.1.

The APN-AMBR is used to limit all bit rates transmitted or received across all default bearers and across all PDN connections of the same

Table 8.1   Subscription and QoS Information Stored at the HSS

Field	Description
IMSI	IMSI is the main reference key used to access the USIM subscription information for a UE
Subscribed-UE-AMBR	Maximum aggregated uplink and downlink MBRs to be shared across all Non-GBR bearers according to the subscription of the user
One or more PDN subscription information:	
PDN address	Indicates subscribed IP address(es)
PDN type	Indicates the subscribed PDN Type (IPv4, IPv6, IPv4v6, Non-IP)
Access Point Name (APN)	A label according to DNS naming conventions describing the access point to the packet data network
EPS subscribed QoS profile	Bearer level QoS parameter values for that APN default bearer (QCI and ARP)
Subscribed-APN-AMBR	Maximum aggregated uplink and downlink MBRs to be shared across all Non-GBR bearers, which are established for this APN

APN. If traffic exceeds this limit, it got discarded. A Single default bearer could utilize the entire APN-AMBR if no other bearer exists. The P-GW enforces the APN-AMBR for both downlink and uplink. In addition, enforcement of APN-AMBR in uplink is done by the UE. Tables 8.2, 8.3, 8.4, and 8.5 show the QoS subscribed information stored at the MME, S-GW, P-GW, and UE, respectively, which is used to enforce QoS parameters and counters [39].

The UE-AMBR is used to limit all bit rates transmitted or received across all PDN connections and across all APNs. That is, the UE-AMBR is set to the sum of the APN-AMBR of all active APNs up to the value of the subscribed UE-AMBR. The UE-AMBR limits the aggregate bit rates transmitted or received across all default non-GBR bearers of a UE. If all traffic of a UE exceeds this counter, it got discarded. A single default bearer could potentially utilize the entire UE-AMBR if no other bearer exists. The network enforces the UE-AMBR in uplink and downlink except for PDN connections using the Control Plane CIoT EPS Optimization.

Table 8.2   QoS Information Stored at the MME

Field	Description
IMSI	IMSI (International Mobile Subscriber Identity) is the subscriber permanent identity
Subscribed UE-AMBR	Maximum aggregated uplink and downlink MBR values to be shared across all Non-GBR bearers according to the subscription of the user
For each active PDN connection:	
APN in Use	APN currently used
PDN type	IPv4, IPv6, IPv4v6, or Non-IP
IP address(es)	IPv4 address and/or IPv6 prefix
EPS subscribed QoS profile	Bearer level QoS parameter values for that APN default bearer (QCI and ARP)
Subscribed APN-AMBR	Maximum aggregated uplink and downlink MBR values to be shared across all Non-GBR bearers, which are established for this APN, according to the subscription of the user
For each EPS bearer within the PDN connection:	
EPS bearer ID	An EPS bearer identity uniquely identifies an EPS bearer for one UE
EPS bearer QoS	QCI and ARP
TFT	Traffic Flow Template
Serving PLMN-Rate-Control	Limits the maximum number of NAS Data PDUs per deci hour sent per direction (uplink/downlink) using the Control Plane CIoT EPS Optimization for a PDN connection

Traffic flows move in both directions: in downlink and uplink and between UE and PDN in both directions. Traffic flows include TCP or UDP traffic and characterized by common traffic parameters for all packets belongs to the same traffic flow such as the same source IP address, destination IP address, source port, destination port, or protocol.

To map a traffic flow to an EPS bearer at both the UE and P-GW, network uses *Traffic Flow Template (TFT)*. A TFT is a set of packet filters that is associated with a specific EPS bearer. It can be defined

Table 8.3   QoS Information Stored at the S-GW

Field	Description
IMSI	IMSI (International Mobile Subscriber Identity) is the subscriber permanent identity
Serving PLMN-rate-control	Limits UE traffic and permits detection of abusive UEs
For each PDN connection:	
APN in use	APN currently used
PDN type	IPv4, IPv6, IPv4v6, or Non-IP
Default bearer	Identifies the default bearer within the PDN connection by its EPS bearer ID
For each EPS bearer within the PDN connection:	
EPS bearer ID	An EPS bearer identity uniquely identifies an EPS bearer for one UE
TFT	Traffic flow template
EPS bearer QoS	QCI and ARP

as a DL TFT or an UL TFT. A TFT contains a number of packet filters that are used to classify packets. A filter contains a number of information to classify packets accordingly. For example, a TFT can contain a number of packet filters where each filter classifies the packets according to a specific source IP address, destination IP address, source port, destination port, or protocol. Packets that match a packet filter are considered to match the TFT of this packet filter. When a traffic flow matches either a DL TFT or an UL TFT, it got routed to the EPS bearer associated with this TFT.

The default EPS bearer can have a DL TFT or an UL TFT associated with it. In such a case, a packet got routed to the default bearer if it matches a packet filter associated with the TFT. If the packet does not match a packet filter, it is dropped. In addition, the default EPS bearer can have no TFT associated with it and in such a case any traffic flow got routed to the default bearer [39].

Downlink packets coming from the Internet or applications and destined for a UE are first received at the P-GW which classify them according to the DL TFTs and if a DL TFT is found to match the traffic flow, P-GW routes the packets to the EPS bearer associated with this DL TFT. If the EPS bearer does not have a DL TFT, P-GW routes those packet to this EPS bearer.

Uplink packets transmitted by the UE to the Internet or applications are received and classified according to the UL TFTs and if an UL TFT

Table 8.4   QoS Information Stored at the PDN Gateway (P-GW)

Field	Description
IMSI	IMSI (International Mobile Subscriber Identity) is the subscriber permanent identity
RAT type	Set to NB-IoT
For each APN:	
APN in use	APN currently used
APN-AMBR	Maximum aggregated uplink and downlink MBR values to be shared across all Non-GBR bearers established for this APN
APN-rate-control	Limits the maximum number of uplink/downlink packets and the maximum number of additional exception report packets per a specific time unit (e.g., minute, hour, day, week) for this APN
For each PDN Connection within the APN:	
IP address(es)	IPv4 address and/or IPv6 prefix
PDN type	IPv4, IPv6, IPv4v6, or Non-IP
Default bearer	Identifies the default bearer within the PDN connection by its EPS bearer ID
Serving PLMN-rate-control	Limits the maximum number of uplink/downlink messages per a specific time unit (e.g., minute, hour, day, week) for a PDN connection.
For each EPS bearer within the PDN connection:	
EPS bearer ID	An EPS bearer identity uniquely identifies an EPS bearer for one UE
TFT	Traffic Flow Template
EPS bearer QoS	ARP, QCI

is found to match the traffic flow, UE transmits the packets on the radio bearer associated with this UL TFT. If the radio bearer does not have an UL TFT, UE routes those packet to this radio bearer.

Figure 8.2 illustrates the radio and EPS bearers established between the UE and eNodeB and between the eNodeB and P-GW, respectively.

Table 8.5   QoS Information Stored at the UE

Field	Description
IMSI	IMSI (International Mobile Subscriber Identity) is the subscriber permanent identity
For each active PDN connection:	
APN in use	APN currently used
APN-AMBR	Maximum aggregated uplink and downlink MBR to be shared across all Non-GBR bearers, which are established for this APN
Assigned PDN type	The PDN Type assigned by the network (IPv4, IPv6, IPv4v6, or Non-IP)
IP address(es)	IPv4 address and/or IPv6 prefix
Default bearer	Identifies the default bearer within the PDN connection by its EPS bearer ID
APN-rate-control	Limits the maximum number of uplink packets and the maximum number of additional exception report packets per a specific time unit (e.g., minute, hour, day, week) for this APN
Serving PLMN-rate-control	Limits the maximum number of uplink NAS Data PDUs per deci hour using the Control Plane CIoT EPS Optimization
For each EPS bearer within the PDN connection:	
EPS bearer ID	An EPS bearer identity uniquely identifies an EPS bearer for one UE
TFT	Traffic Flow Template

The EPS bearers transport data traffic between the UE and Internet as follows:

- At the UE, a traffic flow is mapped to an uplink radio bearer if either an UL TFT matches incoming packets or the radio bearer does not have an UL TFT.

- At the P-GW, a traffic flow is mapped to a downlink EPS bearer if either a DL TFT matches incoming packets or the EPS bearer does not have a DL TFT.

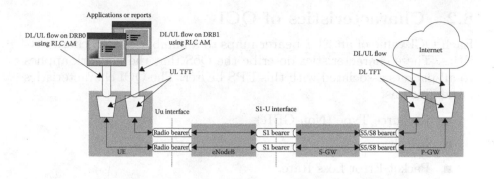

Figure 8.2: End-to-end QoS for EPS bearers.

- A radio bearer, between UE and eNodeB, transports the packets of an EPS bearer. There is a one-to-one mapping between an EPS bearer and this radio bearer.

- An S1 bearer transports the packets of an EPS bearer between an eNodeB and an S-GW.

- An E-RAB is an EPS Radio Access Bearer that refers to the concatenation of an S1 bearer and a radio bearer.

- An S5/S8 bearer transports the packets of an EPS bearer between an S-GW and a P-GW (PDN).

- A UE stores a mapping between an UL TFT and a radio bearer to map a traffic flow into a radio bearer.

- A P-GW stores a mapping between a DL TFT and an S5/S8 bearer to map a traffic flow into an S5/S8 bearer.

- An eNodeB stores a one-to-one mapping between a radio bearer and an S1 bearer to map a radio bearer into an S1 bearer and vice-versa in both the uplink and downlink directions.

- An S-GW stores a one-to-one mapping between an S1 bearer and an S5/S8 bearer to map an S1 bearer into an S5/S8 bearer and vice-versa in both the uplink and downlink.

An NB-IoT device typically has a single radio bearer, single EPS bearer, and connected to a single P-GW.

8.2 Characteristics of QCI

Each QCI value of an EPS bearer maps to a number of QoS characteristics. These characteristics describe the QoS that the network applies to all packets associated with this EPS bearer. The QCI characteristics are defined as:

- Resource Type (Non-GBR).
- Priority.
- Packet Error Loss Rate.

Note that packet delay is not among those characteristics as packet delay cannot be guaranteed for NB-IoT UEs. Those characteristics which apply to each EPS bearer ensure that packets transmitted or received on this EPS bearer receive the same treatment and QoS guarantees. The QCI characteristics are as shown in Table 8.6 [40].

The priority level is interpreted as a lowest priority level value corresponds to the highest priority. The priority level is different from the ARP although both of them are used by the network to prioritize and differentiate traffic in case of network congestion, establishment or modification of EPS bearers. Packet Error Loss Rate (PELR) defines an upper bound of a non-congestion related packet loss rate.

Table 8.6   QCI Characteristics

QCI	Type	Priority Level	Packet Error Loss Rate	Example Services
5		1	10^{-6}	IMS signalling
6		6	10^{-6}	Video (buffered streaming) TCP (e.g., www, e-mail, chat, ftp)
7		7	10^{-3}	Voice, Video live streaming Interactive gaming
8	Non-GBR	8	10^{-6}	Video (buffered streaming)
9		9		TCP (e.g., www, e-mail, chat, ftp)
69		0.5	10^{-6}	Mission critical delay sensitive signalling (e.g., MC signalling and video signalling)
70		5.5	10^{-6}	Mission critical data (same as QCI 6, 8, 9)
79		6.5	10^{-2}	Vehicle-to-Everything messages
80		6.8	10^{-6}	Low latency eMBB applications, Augmented reality

8.3   Quality of Service for UE Using CIoT EPS Optimization

A UE that uses CIoT EPS optimization establishes only SRBs and does not establish any DRB. Traffic carried on these SRB are signalling traffic, such as NAS traffic, between UE and eNodeB. UE can also send small amount of data-plane traffic (e.g., IP, non-IP, SMS, or application-specific traffic) on SRBs by encapsulating them in NAS PDUs that carry the data-plane packets.

The network, through NAS and P-GW, control the bit rate to/from the UE that uses CIoT EPS optimization. The network can uses either of the following two parameters to control the traffic of a UE operating in CIoT optimization:

■ Serving PLMN-Rate-Control: This is used to allow the network to measure and protect network from excessive traffic generated by data traffic carried as part of the NAS signalling. Network sets a limit on the number of data-traffic carried by NAS PDUs that can be sent on DL or UL per deci-hour.

■ APN-Rate-Control: This is used by network to offer a subscribed packet rate expressed as a maximum number of packets to/from a UE per time unit (e.g., day).

Both UE and P-GW comply with the serving PLMN rate control by limiting DL and UL data-plane traffic to the value of this parameter. If DL/UL traffic exceeds this limit, network can discard or delay these packets.

APN-Rate-Control is configured in the P-GW and can be configured in the UE as well. Both UE and P-GW comply with this parameter. This parameter measures only data-plane traffic that go through the APN either for DL or UL. If DL/UL traffic exceeds this limit, network can discard or delay these packets. Moreover, If the traffic exceeds this limit, network can still allow exceptional reports generated by UE to be transmitted.

8.4   QoS Challenges for NB-IoT

NB-IoT devices are physical small devices, with low cost, and limited capabilities for QoS. NB-IoT is not intended for full-fledged QoS applications, such as high-bandwidth video streaming or gaming, instead they only support delay-tolerant application. Lack of full QoS support in NB-IoT devices can be attributed to the following factors:

1. **Limited Data Rate**: NB-IoT can only be assigned a maximum of one resource block on downlink and a maximum of one resource unit on uplink as in Table 7.2. This limits the maximum PHY data rate that can be used to few hundreds kilobytes of bits as shown in Table 7.33. Data rate seen by different applications and services at the application layer can be much reduced due to overhead and transport block repetitions at the PHY sublayer.

2. **Spatial Multiplexing and Transmit Diversity**: A legacy LTE™ device supports transmitting and receiving on multiple antenna for both downlink and uplink (e.g., eight antennas are supported or Massive MIMO technology). This enables the UE to transmit and receives higher volume of packets and more reliably. NB-IoT UE only supports up to two antennas on the downlink for transmit diversity. NB-IoT UE does not support DL neither UL spatial multiplexing.

3. **Half-Duplex FDD**: A legacy LTE can support FDD and TDD for downlink and uplink transmissions. This increase the UE throughput in both directions. NB-IoT UE only supports half-duplex FDD which reduces the maximum throughput and data rate transmitted by the UE.

4. **Channel Quality Indicator (CQI) and Measurement**: CQI are sent by a legacy LTE device on the uplink to the eNodeB to provide a report on the downlink channel quality and conditions. Channel measurements are PHY measurements conducted for handover decisions. While an NB-IoT UE device supports intra-frequency and inter-frequency neighboring cells measurements, NB-IoT UE does not support transmitting CQI neither measure-ments for mobility or handover. This further reduces NB-IoT device data rate and maximum achievable throughput in case of poor channel quality.

5. **Packet Delay Guarantee in Core Network**: Core network does not guarantee packet delay for NB-IoT devices as shown in Table 8.6. Thus, real-time traffic cannot meet its stringent delay bounds.

Chapter 9

Use Cases and Deployment

9.1 NB-IoT Devices

NB-IoT devices are expected to explode to millions of devices. Those devices are collecting a huge amount of structured and unstructured data that are transmitted to a central location (e.g., Cloud infrastructure) where the data get stored, analyzed, and presented to users. NB-IoT devices are typically found in sensors and actuators.

9.1.1 Sensors

A sensor measures, determines, or senses a particular parameter of a system or environment. The sensor reports this parameter in a manner that can be compiled and understood by humans or other devices. Examples of sensors are weather, ambient, or clinical thermometer sensors that sense climate condition, light, or human body heat.

9.1.2 Actuators

An actuator is a special type of device that takes an action based on a system behavior. A sensor reports the status of a particular parameter of a system, whereas an actuator can act to influence that parameter or other parts of the system. An example of an actuator is a pipeline valve that is used in a natural-gas or oil refinery plant where the valve controls the volume of gas or oil flow in the pipeline according to an internal pipeline pressure. The flow valve can also shut down the pipeline and transmit an emergency report if a leak is detected.

9.2 Smart Parking

In recent years, with the increasing number of populations and vehicles, parking and finding an empty parking spot becomes a hurdle and a challenge for many drivers around the world especially in cities with high density of population. The common method of finding a parking spot is a manual method where the driver usually finds a spot in the street through ad-hoc, luck, or experience. If the driver is driving in a city, there may be times when he/she cannot find a parking spot during emergencies. As an alternative, to improve efficiency and usage of fuel and time, there should be a system where the driver can know if there is a parking spot near the destination, prior to his/her arrival at the destination.

The impact of parking pain in major countries has been studied comprehensively in [41]. The time spent and cost of finding an on-street or off-street (underground garage) parking spot is summarized in Table 9.1. The same amount of time is almost spent in searching for on-street or off-street parking. A driver parks on-street approximately 40% of the time with a notable exception being New York City (54%). On average, a driver spends 107 hours a year searching for a parking spot in New York while a driver spends around 65 hours in London or Frankfurt. New York has the highest parking search cost of $2,243 per driver per year. This is due to the annual hours spent searching for a parking spot, at 107 hours in New York. Drivers in the New York waste $4.3 billion per year in time, fuel, and emissions searching for an empty parking spot.

Smart parking for cars, vehicles, and motorcycles can now utilize NB-IoT devices with ultrasonic sensors for detecting the availability of a parking spot. Each sensor has an NB-IoT UE chip. The UE senses the availability of a parking spot, and sends the data through the eNodeB, to a central server (or gateway). The server receives all data from cellular and regional NB-IoT devices and store them in a Cloud-based storage area for further processing and analysis. The server and storage can be both co-located in the Cloud.

When using the NB-IoT for smart parking, each parking spot is set up with a sensor. The sensor node is a small device with extremely low power consumption that consists of an NB-IoT module and ultrasonic device. The sensor node is installed in parking spots. The node can be

Table 9.1 Parking Time and Cost around the World

Country	City	ON-Street Search Time (min/trip)	OFF-Street Search Time (min/trip)	% ON-Street Parking	Annual Search Time (h/Driver/Year)	Cost/Driver /Year	Total/City /Year
United States	New York	15	13	54	107	$2,243	$4.3 bn
	Los Angeles	12	11	49	85	$1,785	$3.7 bn
United Kingdom	London	12	10	44	67	£1,104	£4.3 bn
	Manchester	8	7	39	41	£688	£169 m
Germany	Frankfurt	10	7	42	65	€1,410	€702 m
	Berlin	9	6	48	62	€1,358	€1.8 bn

activated every few seconds. If there is a change in the status, the new status will be sent to the Cloud server. After reporting the status, the node can go to sleep mode.

NB-IoT devices can send full information about the status of each parking spot, time, and date. This information is then shared among all other drivers who subscribe to this service. A driver heading to a downtown area can now receive parking information on his car dashboard informing the driver about the exact locations of empty parking spots at the destination. The driver can book a parking spot up-front or proceed directly to the empty spot. Driver is charged for his/her occupancy or reservation of the parking spot for the calculated amount of time. The description of such a system is best illustrated in Figure 9.1.

With this smart parking system, the driver will be able to save a lot of time, effort, and cost for finding an empty parking spot. The parking officer on the other hand can use the parking information to better manage and handle billing information and offer discounts based on parking utilization, and several other analytic tasks. Other solutions for using NB-IoT sensors for smart parking and payment systems are also possible and proposed [42, 43].

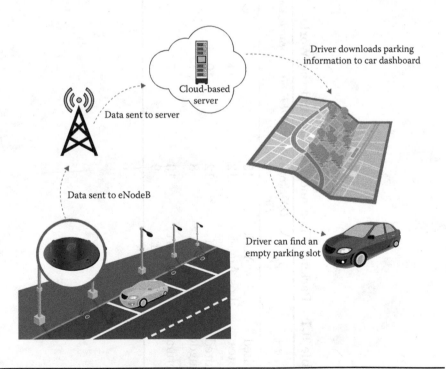

Figure 9.1: Smart parking system.

9.3   Smart City

A smart city is the city with the human, economic, technical, governmental, private, and political capabilities that support unlimited number of technological applications that deliver communal services to residents, citizens, commuters, and businesses. World population continues to flock to urban areas. Smart cities are anticipated to use NB-IoT technology that touches the life of every individual.

Smart city has gained popularity in the recent years because of its benefits to the welfare of its habitants, businesses, and organizations [44]. Smart city contributes to reducing costs, improving efficiency of city resource consumption, engaging more actively with its citizens during emergencies, and reducing man-power and work hours to a great extent. Smart city applications are shown in Figure 9.2.

Modern and smart city will include many NB-IoT applications in the area of energy plants and management, public bus services, and underground transportation, street lights, traffic signals and lights, law enforcement, sewage and water systems, waste management, gas and

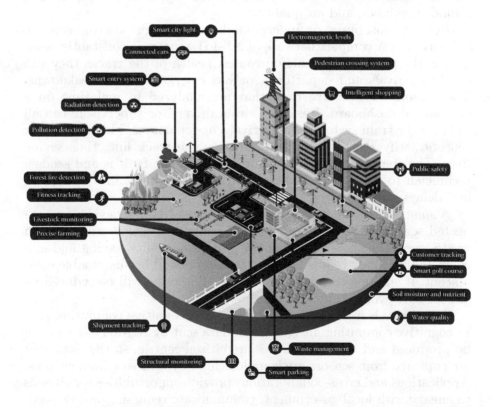

Figure 9.2: Smart city.

electricity metering, and several other applications. Smart city not only implements smart applications but it also fosters a data-driven economy. Smart city benefits not only its residents but also its tourists, investors, and government. These applications rely on using NB-IoT combined with sensors or actuators to transmit a huge amount of structured and unstructured data that can be used for automation, decision-making, and analysis.

Smart electricity grids make electricity delivery more efficient by applying analytics to data that are sent from NB-IoT sensors installed throughout the grid. NB-IoT sensors monitoring the grid are connected and send their data to a Cloud-based server for configuring, controlling, and analyzing the grid. Grid operators can use the data to know and project demand and capacity.

Environmental NB-IoT sensors are used to monitor public waterways, parks, and green spaces. The data sent by these sensors are used to identify spaces that require cleanup or protection. These environmental sensors can also be used to track ambient environmental conditions at different locations throughout the city, such as temperature, humidity, rainfall, and air quality.

Public trains, buses, or underground transit system are connected in a smart city. A complex network of NB-IoT sensors is constantly monitoring the position of trains on routes, health of the tracks they run on, bus arrivals and departures, or bus emergencies or breakdowns. The sensors capture live data that are analyzed in real time on a customized dashboard. Residents, train drivers, or supervisors can all monitor the train system, train arrivals, or departures. Moreover, track sensors notify the operator of a fault on the track line, thus saving lives. The operator can know where and what the fault is and sends a technician to fix it. The fault is fixed, trains go back into operation, and delays are minimized.

A smart waste and sanitization system uses NB-IoT devices for connected sensors. They connect garbage containers which send data on waste levels. The collection vehicles will be more fuel saving and efficient, taking smart routes detected by sensors and be sustainable complacent. If all done correctly, waste collection costs will be reduced by almost 80% over the next decade.

Smart city is a great benefit for applying cognitive computing [45]. In cognitive computing, insights from data sent by NB-IoT sensors can be produced and analyzed. Cities are citizen-centric, so the data that are captured from sensors are complemented by feedback from citizens. Applications and cross-collaborations provide opportunities for citizens to engage with local government, communicate requests, provide feedbacks, or report faults with utilities and infrastructures.

Growth of smart cities is accelerating over the coming few years. Their potential and benefits are limitless. They will touch the lives of many residents by improving living cost, health, and quality of life [46].

9.4   Smart Home

Smart home is one of the fast growing areas for NB-IoT. Voice-assisted devices are becoming the new norm in today homes. NB-IoT is also expected to invade homes with many applications and use cases. NB-IoT devices for home are expected to be installed everywhere: bedrooms, living rooms, kitchen, bathrooms, hallways as in Figure 9.3. Home residents can control their lighting systems from anywhere, receive alerts on their mobile phones when a smoke is detected, monitor their kids, and turn appliances on/off while driving to/from

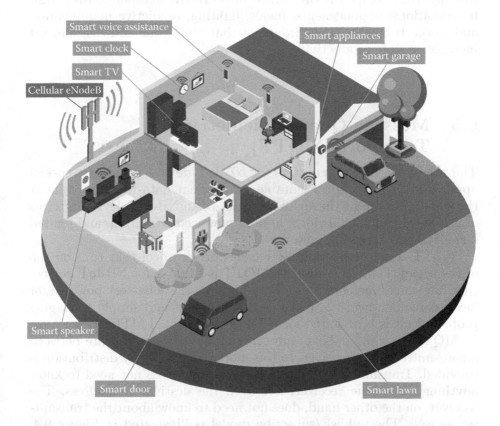

Figure 9.3: Smart home.

work. Using NB-IoT devices in smart home is extremely useful in cases of emergencies like smoke and hazard monitoring or fire alarming.

It is projected that billions of NB-IoT-connected devices for home uses will be reached by 2020. Connected devices can include smart cooking appliances, smart refrigerators, smart air conditioners, smart washing machines, and smart dishwashers. The beauty of connected homes is that smart devices are connected via a central platform which is used to solve many problems. The central platform can be voice-operated control system with artificial intelligence capabilities that help with day-to-day annoyances and routines based on the data collected from the appliances. The control system, based on the data collected by NB-IoT devices, can manage resident's calendar, utilities billing and consumption, and entry and security systems.

Cognitive NB-IoT is a technology that infuses intelligence into home devices, appliances, services, and processes. It is capable of creating a personalized and tailored home environment for its residents during the day. For example, it can advise on workout, weather control, traffic conditions, appointments, meals, lighting, predictive maintenance, and alerts. It creates an environment that saves time and cost and yet increases people productivity.

9.5 Message Queue Telemetry Transport (MQTT)

The use of sensors or actuators for NB-IoT devices requires a special application-layer protocol suitable for efficient data-transfer while at the same time can fit the small footprint of an NB-IoT device in terms of limited channel bandwidth, limited power consumption, and limited memory and processing power.

MQTT is an application-layer transport protocol that runs on top of the legacy TCP/IP protocols. MQTT is suitable for NB-IoT devices that have small memory and processing power, are battery powered, or have scarce bandwidth. MQTT is a lightweight and simple messaging protocol that is best suited for NB-IoT devices and MTC [47].

MQTT uses the publish/subscribe model to communicate between a transmitter and a receiver. In this model, one-to-many distribution is provided. Transmitting applications or devices does not need to know anything about the receiver, not even the destination address. The receiver, on the other hand, does not need to know about the transmitter as well. The publish/subscribe model is illustrated in Figure 9.4. In this figure, a single client publishes its data to the server while

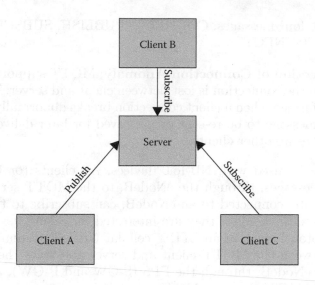

Figure 9.4: Publisher/subscribe model.

other clients may subscribe to the server to receive such data from the publisher.

MQTT minimizes usage of network bandwidth and device resource requirements while at the same time targets to ensure reliability and message delivery. It also possesses the following features [48]:

- **Data-centric**: MQTT transfers data content as byte array. It does not care about content such as HTTP.

- **Application Decoupling**: MQTT transmitters and receivers do not need to be aware of the existence of each other. None of them know the address of others and they just need to care about the content to be delivered or received.

- **Low Protocol Overhead**: MQTT message headers are kept as small as possible. The fixed header is just 2 bytes.

- **Low Bandwidth**: MQTT does not need a high bandwidth. Instead, its on-demand, push-style message distribution keeps network utilization low.

- **Multiple QoS Levels**: MQTT QoS levels allow flexibility in handling different types of messages. MQTT supports delivery of at most once, at least once, or exactly once.

- **Easy of Use**: MQTT uses and implements a small and simple number of command messages. Applications can be implemented using four messages: CONNECT, PUBLISH, SUBSCRIBE, and DISCONNECT.

■ **Detection of Connection Anomaly**: MQTT supports the case when the connection is lost between client and server. The server is informed when a client connection breaks abnormally, allowing the message to be re-sent or preserved for later delivery. Server notifies all other clients.

MQTT can be used with NB-IoT devices. All clients (or UEs) publish their messages, through the eNodeB, to the MQTT server. Other MQTT clients, connected to an eNodeB, can subscribe to the MQTT server to receive the data they are interested in. Figure 9.5 shows the MQTT protocol stack in an LTE™ cellular NB-IoT system. The connection between the MQTT client and server goes over the air interface to the eNodeB, through the EPS (S-GW and P-GW), and finally reaches the MQTT server. MQTT protocol messages are transmitted and received over radio bearer and EPS bearer as explained in Chapter 8.

9.5.1 *Publish/subscribe model*

MQTT protocol is using a publish/subscribe model. The center piece of this model is the use of what is called *topics*. MQTT devices are either a client or a server. A client can publish messages to a topic. A client can also subscribe to a topic that pertains to it and thereby receives any message published to this topic by any other client that publishes to this topic.

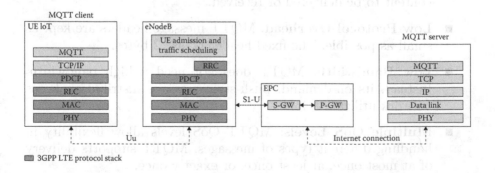

Figure 9.5: MQTT protocol stack.

9.5.2   Topic and subscription

The client in MQTT publishes messages to a topic (or a number of topics). A topic is typically a representation of subject areas. Client can sign up to receive particular messages by subscribing to a topic. Subscriptions can be explicit which limits the messages that are received to the specific topic at hand. Subscriptions can also use wildcard designators, such as a number sign (#), to receive messages for a number of related topics.

9.5.3   Retained messages

MQTT server can keep the message even after sending it to all subscribers. If a new subscription is submitted for the same topic, any retained messages are then sent to the new subscribing client.

9.5.4   Will

When a client connects to a server, it can inform the server that it has a will, or a message, that should be published to a specific topic or topics in the event of an unexpected disconnection. A will is particularly useful in an alarm situation where a user must know immediately when a remote sensor has lost connection with the network.

9.5.5   Quality of service levels

MQTT supports three quality of service (QoS) levels for message delivery to the server. Each level designating a higher level of effort by the server to ensure that the message gets delivered to clients. QoS is symmetric which means that the same QoS level is guaranteed both from client to server and from server to other clients. Higher QoS levels ensure higher reliability for message delivery but can consume more network bandwidth or subject the message to delays due to retransmissions.

MQTT supports three levels of message delivery as follows:

- **QoS Level 0: At most once:** This is a one-way handshake. Sender publishes the message only once. Server can receive the message only once or not at all. No retry is performed by the sender and no response is sent by the server. Message can be lost between sender and server.

- **QoS Level 1: At least once:** This is a two-way handshake. Sender publishes the message and receives an ACK from the server. If an ACK is not received, the sender publishes the message again. Server receives the message and sends an ACK.

If the sender does not receive an ACK from the server, it retransmits the message again until an ACK is received. Message can be duplicated between sender and server.

■ **QoS Level 2: Exactly once**: This is a four-way handshake. Sender publishes the message and receives an ACK from the server. If an ACK is not received, the sender publishes the message again. When the sender receives an ACK, it sends a RELEASE message to the server. The sender expects to receive an ACK for the RELEASE message and if not, it retransmits the RELEASE message. If the server receives a duplicate message, it responds by an ACK and never send the duplicate message again to others.

QoS level 0 is the lowest and loosest QoS while QoS level 2 is the highest and stringent QoS level. UE can choose the QoS that is most appropriate to the application. MQTT-Sensor Network (MQTT-SN) is another variant of MQTT aimed at IoT devices used as sensors. MQTT-SN uses non-IP protocol [49].

9.5.6 MQTT telemetry

MQTT has several uses cases for IoT applications [50]. Its main use, as proposed by IBM, is for telemetry where telemetry is the method used for collecting remote data and send it to a central location for processing and analysis.

With the introduction of the 3GPP™ NB-IoT, it became possible and visible to use MQTT for cellular network. Figure 9.6 shows an example of using MQTT with 3GPP NB-IoT for smart energy application. In this application, three smart meters for three households, implemented as MQTT clients and using NB-IoT, are publishing their meter readings periodically to the MQTT server. Each smart meter publishes its readings to a specific topic. The topic chosen is expressed as an UTF-8 string in the form of "IoT/SmartEnergy/MeterA."

The energy company or the supervisor who is responsible for reading data from the smart meters can do the collection using the MQTT subscription. He can retrieve the smart meter readings by subscribing to the MQTT server. The supervisor subscribes to the same topic but uses a wildcard in the form of "IoT/SmartEnergy/*" to retrieve all readings for all smart meters. The meter readings are now collected, analyzed, and presented in a dashboard to the supervisor who starts analyzing the data, extracts the energy consumption of each household, and averages consumption per city, suburb, or region. Billing and invoices can be subsequently automated and mailed to each household.

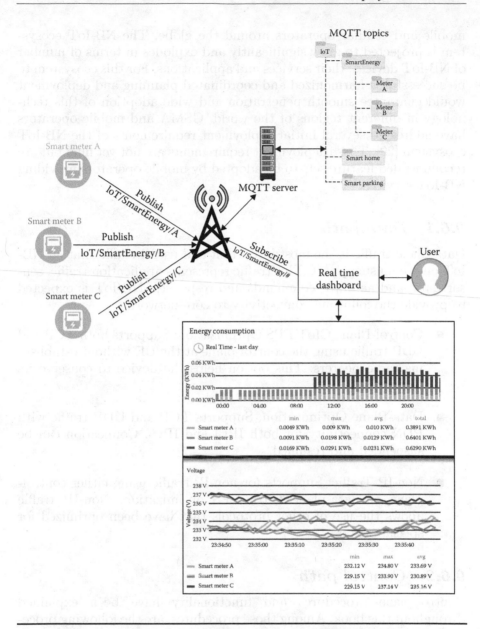

Figure 9.6: Smart metering application using MQTT.

9.6 NB-IoT Baseline Deployment

Cellular NB-IoT is a new and advanced technology that has been standardized in a short amount of time (starting from Release 13 to Release 15). This technology will be administered and operated by mobile and cellular operators around the globe. The NB-IoT ecosystem is projected to grow significantly and explodes in terms of number of NB-IoT devices, their services and applications. For this ecosystem to be successful, a harmonized and coordinated planning and deployment would guarantee smooth penetration and wide adoption of this technology in different regions of the world. GSMA and mobile operators have addressed several initial deployment requirements of the NB-IoT ecosystem [29]. These deployment requirements are not yet final but are recommended feature sets to be adopted by mobile operators providing NB-IoT services.

9.6.1   Data path

Data-plane traffic is the most important and useful traffic that an NB-IoT device must support. This traffic represents application traffic, sensor data, and actuator commands and responses. NB-IoT is expected to provide the following connectivity to core network:

■ Control-Plane CIoT EPS Optimization: Supports both TCP and UDP traffic using the control-plane at the UE without establishing radio bearers. This option helps the device to conserve its battery.

■ Data-Plane Optimization: Supports TCP and UDP traffic with PDN connectivity for both IPv4 and IPv6. Connection can be suspended and resumed.

■ Non-IP Traffic: Supports for non-IP traffic using either control-plane or data-plane CIoT EPS optimization. Non-IP traffic allows the use of other protocols that have been optimized for special purposes.

9.6.2   Control path

Control-plane procedures and functionality have been explained throughout this book. Among those procedures, are the following procedures that NB-IoT devices will support as the basic minimum features to be deployed:

■ Cell Selection and Reselection: ensure that an NB-IoT device is connected to good quality cells.

■ Paging: ensure that the device is signalled while in IDLE mode for any configuration changes or incoming/outgoing calls.

- PSM and DRX: support conserving energy and longer battery life time while in IDLE or CONNECTED modes.

- Control-Plane CIoT Optimization: Supports IP or non-IP traffic while minimizing energy consumption and protocol stack configurations and operations.

9.6.3 Deployment bands

NB-IoT is expected to deployed worldwide. 5G has identified several bands to be used with NB-IoT services as indicated in Table 7.3. However, some initial bands are popular to be used for NB-IoT deployments. This is shown in Table 9.2.

9.6.4 Deployment modes

NB-IoT devices are expected to be deployed using the three modes of operation: inband, guardband, and standalone as in Figure 2.12. However, guardband deployment is attractive as it does not consume the LTE band thus does not affect network capacity or network resources. Instead, it utilizes unused bands while increasing network and spectrum utilization.

9.6.5 Coverage enhancement

NB-IoT applications and devices are expected to be deployed in areas where cellular coverage does not usually have a good signal propagation and penetration such as underground parking, garages, elevators,

Table 9.2   Initial Deployment Bands for NB-IoT

Region	Band
North America	B13 (700) B4 (1700)
Europe	B3 (1800), B8 (900),B20 (800)
Asia Pacific	B1 (2100), B3 (1800), B5 (850), B8 (900), B18 (850), B20 (800), B26 (850), B28 (700)
Latin America	B2 (1900), B3 (1800), B5 (850), B28 (700)
Commonwealth of Independent States	B3 (1800), B8 (900), B20 (800)
Africa	B3 (1800), B8 (900)
Middle East and North Africa	B8 (900), B20 (800)

and ground pits. An NB-IoT device supports coverage enhancements, which increases the signal range and radio coverage. 5G NB-IoT device supports three Coverage Enhancement Levels (CELs): CEL0 for 0 dB, CEL1 for 10 dB, and CEL2 for 20 dB. The selection of a CEL is explained in Section 6.3.2.

Coverage enhancements rely on increasing the transmitted power of signalling channels combined with repeated transmissions which improve the reliability of the message reception by the receiver. However, coverage enhancement can consume additional battery energy.

9.6.6   Power class

NB-IoT supports three power classes that range from conservative to normal in terms of battery power consumption. Power class 3 (23 dBm) is the legacy LTE device power class. Power class 5 (20 dBm) is another power class for NB-IoT devices, which consumes less power. Finally, power class 6 (14 dBm) is an additional power class that consumes even less power.

9.6.7   Quality of service

Chapter 8 explained QoS that can be offered by the core network for NB-IoT devices. NB-IoT devices are expected to be large in number and density and hence their traffic might overwhelm the network and cause congestion. It is expected that the core network shapes the traffic generated by these devices. Such shaping can be performed through QoS rate control metrics: Serving PLMN Rate Control and APN Rate Control.

9.7   Mobile Operator Deployment

In 2017, T-Mobile, a mobile operator in the United States, has launched an initial NB-IoT network that was tested in US cities as part of a smart city project. T-Mobile continues its US-nationwide deployment of NB-IoT throughout the year 2018 [51]. T-Mobile emphasizes the efficiency of NB-IoT network and devices along with the low cost [52]. T-Mobile plans to operate NB-IoT in guardband. Operating NB-IoT in guardband acts as driving on the shoulders of a highway and provides better new services to customers while increasing the mobile operator revenue stream. NB-IoT devices, deployed in guardband, carry data with high efficiency and performance and at the same time, do not compete with normal cellular data or voice traffic.

Verizon, a mobile operator in the United States, offers LTE machine-type services such as wearables, fleet, and asset management. Verizon is planning to deploy an USA-nationwide NB-IoT guardband network before the end of the year 2018.

Mobile operators around the world are continuing the effort of deploying NB-IoT during the year 2018. The NB-IoT service offers customers with large number of usage scenarios and widely varying sets of network capacity and coverage need. NB-IoT provides mobile operators with wide-area coverage networks and low-power devices. NB-IoT offers a peak downlink throughput of few hundreds Kbps while having low power consumption and usage of a small channel bandwidth. It is expected that mobile operators charge about 3–6 USD for each NB-IoT device per year to connect to the NB-IoT network [51, 52].

References

[1] About mobile technology and IMT-2000. [Online]. Available: https://www.itu.int/osg/spu/imt-2000/technology.html

[2] *Overall description; Stage 2*, 3GPP Std. 36.300, Dec. 2017, v15.0.0.

[3] "Requirements related to technical performance for IMT-advanced radio interface(s)," ITU, Tech. Rep. M.2134, Nov. 2008.

[4] "Framework and overall objectives of the future development of IMT for 2020 and beyond," ITU, Tech. Rep. M.2083-0, Sep. 2015.

[5] Y. P. E. Wang, X. Lin, A. Adhikary, A. Grovlen, Y. Sui, Y. Blankenship, J. Bergman, and H. S. Razaghi, "A primer on 3GPP narrowband internet of things," *IEEE Communications Magazine*, vol. 55, no. 3, pp. 117–123, Mar. 2017.

[6] M. Condoluci, G. Araniti, T. Mahmoodi, and M. Dohler, "Enabling the IoT machine age with 5G: Machine-type multicast services for innovative real-time applications," *IEEE Access*, vol. 4, pp. 5555–5569, 2016.

[7] "Electromagnetic compatibility and radio spectrum matters (ERM); system reference document (SRdoc): Spectrum requirements for short range device, metropolitan mesh machine networks (M3N) and smart metering (SM) applications," ETSI, Tech. Rep. 103 055, Sep. 2011, v1.1.1.

[8] *Architecture description*, 3GPP Std. 36.401, Dec. 2017, v15.0.0.

[9] "3GPP low power wide area," White Paper, GSMA, Oct. 2016.

[10] "The next generation of communication networks and services," 5G Vision, 5G Infrastructure Public Private Partnership (PPP), 2015. [Online]. Available: https://5g-ppp.eu/wp-content/uploads/2015/02/5G-Vision-Brochure-v1.pdf.

[11] "Study on provision of low-cost machine-type communications (MTC) user equipment (UEs) based on LTE," 3GPP, Tech. Rep. 36.888, June 2013, v12.0.0.

[12] "Cellular system support for ultra-low complexity and low throughput Internet of Things (CIoT)," 3GPP, Tech. Rep. 45.820, Nov. 2015, v13.1.0.

[13] "Feasibility study on new services and markets technology enablers for massive internet of things; stage 1," 3GPP, Tech. Rep. 22.861, Sep. 2016, v14.1.0.

[14] "Feasibility study on new services and markets technology enablers; stage 1," 3GPP, Tech. Rep. 22.891, Sep. 2016, v14.2.0.

[15] *Radio Resource Control (RRC) protocol specification*, 3GPP Std. 36.331, Jan. 2018, v15.0.1.

[16] "Information technology-ASN.1 encoding rules: Specification of packed encoding rules (PER)," Recommendation ITU-T X.691, Aug. 2015.

[17] *User Equipment (UE) procedures in idle mode*, 3GPP Std. 36.304, Dec. 2017, v14.5.0.

[18] *Physical layer measurements*, 3GPP Std. 36.214, Jan. 2018, v15.0.1.

[19] *UE Radio Access Capabilities*, 3GPP Std. 36.306, Dec. 2017, v14.5.0.

[20] *Non-Access-Stratum (NAS) protocol for Evolved Packet System (EPS); Stage 3*, 3GPP Std. 24.301, Sep. 2017, v15.0.1.

[21] *Mobile radio interface Layer 3 specification; Core network protocols; Stage 3*, 3GPP Std. 24.008, Dec. 2017, v15.1.0.

[22] "Next generation mobile networks," 5G White Paper, NGMN Alliance, 2015.

[23] *Packet Data Convergence Protocol (PDCP) specification*, 3GPP Std. 36.323, Dec. 2017, v14.5.0.

[24] *Security Architecture*, 3GPP Std. 33.401, Jan. 2018, v15.2.0.

[25] *Radio Link Control (RLC) protocol specification*, 3GPP Std. 36.322, Dec. 2017, v15.0.0.

[26] *Medium Access Control (MAC) protocol specification*, 3GPP Std. 36.321, Dec. 2017, v15.0.0.

[27] *Requirements for support of radio resource management*, 3GPP Std. 36.133, Dec. 2017, v15.1.0.

[28] *Physical layer; General description*, 3GPP Std. 36.201, Mar. 2017, v14.1.0.

[29] "NB-IoT deployment guide to basic feature set requirements" 5G White Paper, GSMA, 2017.

[30] "NB-IoT: technical report for BS and UE radio transmission and reception," 3GPP, Tech. Rep. 36.802, July 2016, v13.0.0.

[31] *Base Station (BS) Radio Transmission and Reception*, 3GPP Std. 36.104, Dec. 2017, v15.1.0.

[32] *User Equipment (UE) radio transmission and reception*, 3GPP Std. 36.101, Dec. 2017, v15.1.0.

[33] *Physical channels and modulation*, 3GPP Std. 36.211, Dec. 2017, v15.0.0.

[34] *Multiplexing and channel coding*, 3GPP Std. 36.212, Jan. 2018, v15.0.1.

[35] *Physical layer procedures*, 3GPP Std. 36.213, Dec. 2017, v15.0.0.

[36] X. Lin, A. Adhikary, and Y. P. E. Wang, "Random access preamble design and detection for 3GPP narrowband IoT systems," *IEEE Wireless Communications Letters*, vol. 5, no. 6, pp. 640–643, Dec. 2016.

[37] M. Centenaro, L. Vangelista, S. Saur, A. Weber, and V. Braun, "Comparison of collision-free and contention-based radio access protocols for the internet of things," *IEEE Transactions on Communications*, vol. 65, no. 9, pp. 3832–3846, Sep. 2017.

[38] R. Harwahyu, R. G. Cheng, C. H. Wei, and R. F. Sari, "Optimization of random access channel in NB-IoT," *IEEE Internet of Things Journal*, vol. 5, no. 1, pp. 391–402, Feb. 2018.

[39] *General Packet Radio Service (GPRS) Enhancements for E-UTRAN Access*, 3GPP Std. 23.401, Dec. 2017, v15.2.0.

[40] *Policy and Charging Control Architecture*, 3GPP Std. 23.203, Dec. 2017, v15.1.0.

[41] G. Cookson and B. Pishue. (2017, July) The impact of parking pain in the US, UK and Germany. INRIX Research. [Online]. Available: http://inrix.com/.

[42] B. M. Mahendra, S. Sonoli, N. Bhat, Raju, and T. Raghu, "IoT based sensor enabled smart car parking for advanced driver assistance system," in *2017 2nd IEEE International Conference on Recent Trends in Electronics, Information Communication Technology (RTEICT)*, Bangalore, May 2017, pp. 2188–2193.

[43] J. Shi, L. Jin, J. Li, and Z. Fang, "A smart parking system based on NB-IoT and third-party payment platform," in *2017 17th International Symposium on Communications and Information Technologies (ISCIT)*, Cairns, Sep. 2017, pp. 1–5.

[44] IoT for smart city. [Online]. Available: https://www.microsoft.com/en-us/internet-of-things/smart-city.

[45] M. K. Patra, "An architecture model for smart city using cognitive internet of things (CIoT)," in *2017 Second International Conference on Electrical, Computer and Communication Technologies (ICECCT)*, Coimbatore, Feb. 2017, pp. 1–6.

[46] E. Patti and A. Acquaviva, "IoT platform for smart cities: Requirements and implementation case studies," in *2016 IEEE 2nd International Forum on Research and Technologies for Society and Industry Leveraging a Better Tomorrow (RTSI)*, Bologna, Sep. 2016, pp. 1–6.

[47] MQTT. [Online]. Available: http://MQTT.ORG.

[48] OASIS, "MQTT 3.1.1 specification," Dec. 2015.

[49] A. H. L. T. Stanford-Clark, "MQTT for sensor networks (MQTT-SN) protocol specification version 1.2," Nov. 2013.

[50] IBM. (2018). IBM MQ version 8.0 documentation. [Online]. Available: https://www.ibm.com/support/knowledgecenter/en/SSFKSJ_8.0.0.

[51] T-Mobile, USA. [Online]. Available: https://newsroom.t-mobile.com/news-and-blogs/narrowband-iot.htm.

[52] T-Mobile, USA. [Online]. Available: http://iot.t-mobile.com/.

Index

Uplink data transmission,
134–135
Uplink information transfer,
59–60
Uplink physical
channels/structure
demodulation reference
signal, 193–195
narrowband physical
random access
channel, 192–193
narrowband physical
uplink shared channel,
187–192
power control, 195
resource grid, 185–187
transmission scheme single-
carrier-frequency
division multiple
access, 183–185

User equipment, 123
aggregate maximum bit
rate, 201–202
capability transfer, 60–61

V
Voice-centric communication, 1

W
WAN solution. *See* Wide Area
Network (WAN)
solution
Wide Area Network (WAN)
solution, 9–10
Wireless communication
systems, 1